THE PRICE OF LINGUISTIC PRODUCTIVITY

THE PRICE OF LINGUISTIC PRODUCTIVITY
How Children Learn to Break the Rules of Language

Charles Yang

The MIT Press
Cambridge, Massachusetts
London, England

© 2016 Massachusetts Institute of Technology

All rights reserved. No part of this book may be reproduced in any form or by any electronic or mechanical means (including photocopying, recording, or information storage and retrieval) without permission in writing from the publisher.

This book was set in Times Roman and Mathtime Pro 2 by the author.

Printed and bound in the United States of America.

Library of Congress Cataloging-in-Publication Data

Names: Yang, Charles D., author.
Title: The price of linguistic productivity : how children learn to break the rules of language / Charles Yang.
Description: Cambridge, MA : The MIT Press, [2016] | Includes bibliographical references and index.
Identifiers: LCCN 2016010715 | ISBN 9780262035323 (hardcover : alk. paper)
Subjects: LCSH: Language acquisition–Age factors. | Linguistic analysis (Linguistics) | Bilingualism in children. | Computational linguistics.
Classification: LCC P118.65 .Y36 2016 | DDC 401/.93–dc23 LC record available at https://lccn.loc.gov/2016010715

10 9 8 7 6 5 4 3 2 1

For Vivian Jin Legate-Yang

Contents

Acknowledgments		xi
1	**Border Wars**	1
	1.1 How Grammars Leak	2
	1.2 Where Core Meets Periphery	5
	1.3 Some Outstanding Problems	9
2	**The Inevitability of Rules**	15
	2.1 Statistical Profiles of Grammar	15
	2.1.1 The Long Tail	17
	2.1.2 Quantifying Sparsity	21
	2.1.3 Distributional Sparsity and Language Learning	22
	2.2 Interlude: Irregular Rules and Irregular Verbs	26
	2.3 Productivity in Child Language	31
	2.3.1 The Wug Test	31
	2.3.2 Regularization vs. Irregularization in English	33
	2.3.3 Productivity across Languages	34
3	**The Tipping Point**	41
	3.1 Learning by Generalization	41
	3.2 The Cost of Exceptions	45
	3.3 Elsewhere in Language Processing	49
	3.3.1 Listing Exceptions	50
	3.3.2 Exceptions before Rules	52
	3.4 The Tolerance Principle	60
	3.5 Remarks	66
	3.5.1 Smaller Is Better	66
	3.5.2 Types, Tokens, and Artificial Languages	67
	3.5.3 Effective Vocabulary and Variation	69
	3.5.4 Recursive Tolerance and Structured Rules	71
	3.5.5 Structures and Statistics	75
4	**Signal and Noise**	79
	4.1 When Felt Becomes Feeled	81
	4.1.1 Evaluating Irregulars	83
	4.1.2 Adam, Eve, and Abe	87
	4.1.3 Why Are Noun Plurals Easier to Learn?	91
	4.2 A Recursive Approach to Stress	93
	4.2.1 A Sketch of English Stress	94

		4.2.2	Prosodic Development and Learnability	96

        4.2.2   Prosodic Development and Learnability   96
        4.2.3   Stage Transitions in Stress Acquisition   99
   4.3   The Mysteries of Nominalization   106
        4.3.1   Productivity and Frequency   107
        4.3.2   Form Proposes, Meaning Disposes, Tolerance Decides   109
        4.3.3   Evaluating Nominalization Suffixes   112
   4.4   The Horrors of German: Exceptions that Force the Rules   121
        4.4.1   More Regularity After All   121
        4.4.2   How to Find Subregularities   125
        4.4.3   German Plurals: A Recap   133

5   When Language Fails   139
   5.1   Finding Gaps   140
        5.1.1   Stride, Strode, *Stridden   143
        5.1.2   Stem Alternations in Spanish   147
        5.1.3   Defective Inflections in Russian   152
        5.1.4   The Indeterminacy of Polish Masculine Genitives   154
   5.2   The Rise and Fall of Productivity   157
   5.3   Diagnosing Sickness   160
        5.3.1   The Symptoms   160
        5.3.2   Predicting Case Substitution   165
        5.3.3   The Actuation of Change   166

6   The Logic of Evidence   171
   6.1   Inference and Weight of Evidence   172
        6.1.1   Indeterminacy   172
        6.1.2   Indirect Negative Evidence   174
        6.1.3   Sufficiency of Positive Evidence   177
   6.2   Why Are There No Asleep Cats?   180
        6.2.1   The Failure of Indirect Negative Evidence   181
        6.2.2   *A*-adjectives Are Not Atypical   184
        6.2.3   Generalization with Sufficient Evidence   187
   6.3   Resolving Baker's Paradox   190
        6.3.1   Conditions on Dative Constructions   191
        6.3.2   How to Text Me a Message   198
        6.3.3   Beyond the First Years of Life   206

7	**On Language Design**	215
	7.1 Computational Efficiency in Language Acquisition	215
	7.2 Core and Periphery Revisited	218
	7.3 The Ecology of Language Learning	223
Bibliography		229
Index		257

Acknowledgments

A famous Chinese proverb roughly translates into English as "It takes ten years to forge a good sword" (not "cast sword decade" as Google Translate has it). This project has certainly taken its time. The original goal was modest: How does a child learn that "add -*d*" is the productive rule for English past tense because it readily applies to novel verbs (again, *google* and *googled*), whereas the irregulars patterns — *catch-caught, know-knew, sleep-slept* — are never extended beyond a fixed list? (As I recall, it may have been prompted by a question from Steve Anderson after my first job talk.) With the help of Sam Gutmann, a mathematical solution was quickly found but it was always difficult to sell an equation. A solid collection of case studies had to be gathered, and only very slowly did I grasp the range of problems falling under the scope of the equation.

During the course of this project, I have needed, and indeed received, plenty of help from many friends. Part of section 2.1, a review of the statistics of morphology, is based on joint work with Constantine Lignos. The acquisition of English metrical stress (section 4.2) grew out of an earlier collaboration with Julie Anne Legate. Sarah Murphy helped with the distributional analysis of the English nominalization suffixes (section 4.3). The study of German noun plurals (section 4.4) relied on Beatrice Santorini and Florian Schwarz's native-speaker judgments as well as their linguistic insights. Kyle Gorman, Margaret Borowczyk, and Jennifer Preys collaborated with me on the study of paradigmatic gaps in Spanish, Polish, and Russian (chapter 5). Iris Edda Nowenstein and Einar Freyr Sigurðsson were instrumental as I produced the account of morphosyntactic changes in Icelandic (Section 5.3). I am grateful to them all.

A long gestation period means that this work has benefited from advice and suggestions from many colleagues over the years, including Steve Anderson, Mark Aronoff, Luc Baronian, Gene Buckley, Andrea Ceolin, Harald Clahsen, Elan Dresher, David Embick, John Frampton, Janet Dean Fodor, Amy Forsyth, Lila Gleitman, Sam Gutmann, John Halle, Norbert Hornstein, Ava Irani, Steve Isard, Aravind Joshi, Tony Kroch, Kyle Latack, Mark Liberman, Mitch Marcus, James Myers, Elissa Newport, Lisa Pearl, David Pesetsky, Don Ringe, Tom Roeper, William Sakas, Kathryn Schuler, Mark Steedman, Virginia Valian, and the reviewers for the manuscript. I am especially indebted to Bill Labov and Gillian Sankoff for teaching me sociolinguistics, which help broaden my perspectives on language acquisition and change. Abby Cohn gave the final manuscript a close reading and made numerous suggestions for improvement. Thanks are also due to Marc Lowenthal, Marcy Ross, Sarah Courtney, Amy Hendrickson, and the MIT Press for their support.

This work has been inspired by my teachers, Bob Berwick, Noam Chomsky, and Morris Halle. Erwin Chan, Kyle Gorman, Constantine Lignos, and Qiuye Zhao deserve special thanks for helping me understand language and computation: as can be seen in the following pages, they have taught me much more than I could have taught them. Finally, I thank Julie Anne Legate for many helpful and patient discussions.

Swarthmore, Pennsylvania
December 2015

1 Border Wars

It seems preposterous to write a whole book about an equation (see (3) in section 1.2). So let me begin by laying out the problems that the equation is designed to solve, how these problems are important, and why the solution must be in the form of an equation.

There is no doubt that human language evolved as a biological capacity. It is also almost certainly the case that language emerged relatively recently and suddenly: perhaps no more than 100,000 years ago (Tattersall 2012), a blink of the eye in evolutionary terms. This puts a premium on *Darwin's Problem* (Bolhuis et al. 2014; Hornstein 2009): How to situate language, from the perspective of behavior, brain, and evolution, within the human cognitive and perceptual system, which must be shared in part with other species and lineages?

Speculations abound within and without generative linguistics; see Hauser et al. 2014 for a critical assessment of the current literature. But no amount of evolutionary musing should distract us from the more traditional, and much more tangible, goals of language sciences. A theory of language needs to be sufficiently elastic to account for the complex patterns in the world's languages but at the same time sufficiently restrictive so as to guide children toward successful language acquisition in a few short years (Chomsky 1965). Only then does Darwin's Problem arise: as a statement of human biology, a theory of language can only include evolutionary innovations that would have been plausible in the extremely brief history of *Homo sapiens*.

The current project deals with the boundary issues between language and cognition with an eye on evolution. In one sense, it is the continuation of earlier research. The variational approach to language (Yang 2002, 2004) suggests that children use general learning mechanisms to navigate within the hypothesis space provided by Universal Grammar (UG); in doing so, the variational approach dispenses with domain-specific learning models long known to be problematic. Along a similar line, the present work develops a theory of linguistic representation, learning, and use that, once again, shifts the explanatory burden from Universal Grammar to factors external to language — with specific considerations of computational complexity. In another sense, however, the current study provides an amendment to the variational framework. It investigates the extent to which key properties of language, rather than being built in, can be attributed to children's ability to derive generalizations from the linguistic data. A reduced load for the genetic endowment of language promises a more viable solution to Darwin's Problem but will inevitably exacerbate Plato's Problem (Chomsky 1986): How do children acquire their knowledge of language, which is grossly underdetermined by experience? The answer is partially provided by Universal Grammar, but we cannot be asking for too much — at least no more than what evolution could conceivably offer in the very recent past.

Let's go straight to the heart of matter: the recursive composition of hierarchical structures ("Merge"; Chomsky 1995), representing a giant leap forward in the evolution of language and cognition (Hauser et al. 2002). What are the behavioral, and ultimately evolutionary, benefits of a combinatorial system over a finite inventory of fixed expressions, especially because we now know about the brain's enormous capacity for storage? How does a simple and elegant computational procedure square with the manifest arbitrariness and idiosyncrasies across languages? The Société de Linguistique de Paris once issued a moratorium on theories of language origins; indeed, a credible account of language and its place in cognition and evolution must be grounded firmly in empirical materials. As it happens, our evolutionary reflections turn up some old and unsettled scores in the study of language. And that's well and good. A minimalist UG is only convincing if it engages with, and provides convincing solutions to, the everyday problems that concern working linguists: Le biologiste passe, la grenouille reste.

1.1 How Grammars Leak

As Edward Sapir once famously noted, all grammars leak. In less colorful terms, all languages have exceptions that exist side by side with overarching rules and regularities. But lest this banal observation overshadow Sapir's main message: without a grammar, there will be no leaks to plug.

> It is obvious that a language cannot go beyond a certain point in this randomness. Many languages go incredibly far in this respect, it is true, but linguistic history shows conclusively that sooner or later the less frequently occurring associations are ironed out at the expense of the more vital ones. In other words, all languages have an inherent tendency to economy of expression. Were this tendency entirely inoperative, there would be no grammar. The fact of grammar, a universal trait of language, is simply a generalized expression of the feeling that analogous concepts and relations are most conveniently symbolized in analogous forms. Were a language ever completely grammatical, it would be a perfect engine of conceptual expression. Unfortunately, or luckily, no language is tyrannically consistent. All grammars leak. (Sapir 1928, 38–39)

The fact of exceptions, then, should not deter linguists from formulating theories about the systematic properties of language. When we evoke labels such as *diacritics*, *irregularity*, and *lexicalization* — which can be found in every theorist's toolkit — we are simultaneously committing to a grammar, one associated with *basic* word order, *unmarked* forms, and *default* rules:

> It is quite obvious that many of the phonological rules of the language will have certain exceptions which, from the point of view of the synchronic description, will be quite arbitrary. This is no more surprising than the fact that there exist strong verbs or irregular plurals. Phonology, being essentially a finite system, can tolerate some lack of regularity (exceptions can be memorized); being a highly intricate system, resulting (very strikingly, in a language like English) from diverse and interwoven historical processes, it is to be expected that a margin of irregularity will persist in almost every aspect of the phonological description. Clearly, we must design our linguistic theory in such a way that the existence of exceptions does not prevent the systematic formulation of those regularities that remain. (Chomsky and Halle 1968, 172)

Not everyone agrees. Language scientists and engineers have been wrestling with leaky grammars ever since Sapir. If exceptions are idiosyncratic and must be somehow committed to memory, why not relegate all of language to storage, because the brain is capable of retaining vast quantities of information? To wit, the so-called past-tense debate, to which I return repeatedly in the following pages, has been a struggle over whether *some* verbs (the irregulars; Pinker and Ullman 2002) or *all* verbs (the regulars as well; McClelland and Patterson 2002) are organized as associative memory: the latter position would dispense with Sapir's "symbolized" rules altogether.

The controversy over exceptions intensified further when generative grammar moved from language-specific rules and constructions to universal principles and constraints (Chomsky 1981). If exceptions already pose a serious challenge to the study of particular grammars, how do they figure into a theory that commits to an innate, universal, and invariant predisposition for language?

> It is reasonable to suppose that UG determines a set of core grammars and that what is actually represented in the mind of an individual even under the idealization of a homogeneous speech community would be a core grammar with a periphery of marked elements and constructions. ... Viewed against the reality of what a particular person may have inside his head, core grammar is an idealization. From another point of view, what a particular person has inside his head is an artifact resulting from the interplay of many idiosyncratic factors, as contrasted

with the most significant reality of UG (an element of shared biological endowment) and the core grammar (one of the systems derived by fixing the parameters of UG in one of the permitted ways). (Chomsky 1981, 8)

Exceptions may have been put in their proper place—the periphery—but they have not exactly gone away. Chomsky's formulation brings into focus the problem of language acquisition. When constructing a theory of grammar, linguists have at their disposal a plethora of tools to disentangle the core from the periphery: grammaticality judgments, corpus statistics, historical documents, and an ever-expanding arsenal of experimental methods. And they still disagree over the proper partitioning. How does a young child steer clear of the peripheral idiosyncrasies to acquire a core grammar, all in a few short years? Exceptions are defined in opposition to the grammar, but the acquisition data does not arrive wearing "core" or "periphery" on its sleeves: the learner seems to have a perfect chicken-and-egg problem.

The core-vs.-periphery problem was very much the focus of learnability research in the 1980s. I will review this work in the following pages but it is fair to say that no widely accepted solution has been produced. Sag (2010, 487) summarizes the state of the affairs from the perspectives of a skeptic:

> How are we to know which phenomena belong to the core and which to the periphery? The literature offers no principled criteria for distinguishing the two, despite the obvious danger that without such criteria, the distinction seems both arbitrary and subjective. The bifurcation hence places the field at serious risk of developing a theory of language that is either vacuous or else rife with analyses that are either insufficiently general or otherwise empirically flawed. There is the further danger that grammatical theories developed on the basis of "core" phenomena may be falsified only by examining data from the periphery—data that falls outside the domain of active inquiry.

A possible course to follow is to abandon the core-vs.-periphery distinction. There is a detectable continuity from Gross's taxonomy of French verbs (Gross 1975, 1979) to Sag's radically lexicalized treatment of movement dependencies (2010), from Lakoff's irregular syntax (1970) to present-day Construction Grammars, a network of "stored pairings of form and function" that constitutes the totality of linguistic knowledge (Goldberg 2003, 219). Similarly, according to the usage-based approach to language acquisition, children do not make use of a systematic grammar; rather, "they sometimes have a set expression readily available and so they simply retrieve that expression from their stored linguistic experience" (Tomasello 2000b, 77). In a wide-ranging study, Culicover (1999) investigates numerous syntactic constructions that cannot be attributed to the core parameter system. (Or shouldn't be, for that would require an enormous number of parameters, defeating the very purpose of parameters as compact descriptions of disparate phenomena.) He proposes that language acquisition follows inductive methods, where learners draw generalizations over the entire range of language data. No dichotomy between the core and the periphery is supposed, and child learners have no chicken-and-egg dilemma.

I for one am not quite ready to give up the core. Formal results have consistently shown that a constrained hypothesis space remains the most promising solution to the general problem of learning, of which language acquisition is a special case (Sakas and Fodor 2012; Sakas et al. 2016; Valiant 1984; Vapnik 2000). Additionally, when children's language deviates from the input, it nevertheless remains in a restrictive range of possibilities (Crain 1991; Yang 2002), which further supports the conception of the core as a highly structured system. As will be reviewed in chapter 2, computational and quantitative studies of language suggest that the role of linguistic storage has been greatly exaggerated, and that there is clearcut evidence from child language for a categorical distinction between rules and exceptions, and between the core and the periphery.

But on the resolution of the boundary dispute, I am in agreement with the critics. It is no longer advisable to dodge the question. While some of the purported peripheral idiosyncrasies might only be apparent, it is no longer sufficient to point out how a coreless approach misses important empirical generalizations, or fails to provide a plausible solution to the problem of acquisition. Since not all aspects of language are plausibly innate—the "add -*d*" rule for English past tense, for instance—some kind of data-driven inductive learning is absolutely necessary. A positive answer must be given so that the boundary between the core and the periphery can be drawn, at least for theorists who would like to maintain such a distinction.

1.2 Where Core Meets Periphery

Like most researchers, I started at the core only to be driven to the periphery. In Yang 2002, I developed the variational learning framework for language acquisition and change. The variational model holds language learning to be a probabilistic process: the child has a statistical distribution over the space of possible grammars (or parameter values), and it is this distribution that changes in response to linguistic data. As learning proceeds, the child will access the target grammar with increasing probability, while the nontarget but linguistically possible grammars may still be used, albeit with decreasing probabilities. The competition scheme results in children's occasional but systematic deviation from the target grammar, which will be left standing in the end.

In many ways, the variational learning model is an improvement over traditional *transformational learning* models of which the triggering learning algorithm (Gibson and Wexler 1994) is the paradigm example. Under the transformational scheme, the learner is identified with a single grammar in the hypothesis space (Berwick 1985; Chomsky 1965; Wexler and Culicover 1980). The current grammar is abandoned if it fails to analyze an input utterance, and a new grammar is adopted instead. As pointed out by many researchers of child language (e.g., Bloom 1990; Niyogi and Berwick 1996; Randall 1990; Valian 1991), the triggering model is vulnerable to noise: after years of patient navigation, the learner's grammar may be undone by a single ungrammatical utterance. The variational model, which regards learning as probabilistic, can robustly countenance a certain level of noise. Instead of having a probability of 1 for a parameter value, the learner may settle on 0.99, reserving a noisy margin of 1%.

Variational learning is well equipped to handle noise — and only noise. It does not have the appropriate mechanism for distinguishing noise from exceptions. To take a concrete example, English ceased to be a verb raising language in the Middle English period (Ellegård 1953; Kroch 1989) and now employs periphrastic auxiliaries in question formation. Yet the primary linguistic data does contain instances of main verb raising. (1) is a familiar nursery rhyme, which dates back to the 1700s, when the loss of verb raising was already near completion.

(1) Baa baa black sheep have$_t$ you ___$_t$ any wool?

Suppose (1) and similar sentences appear in 1% of the utterances that a child receives. The variational model is straightforwardly applicable, except it will get the facts wrong. The child will converge on a stable combination that raises the main verb 1% of the time as in (1) while using an auxiliary for the rest. But this is the correct numerical distribution but a wrong structural one. Unless the learner identifies that (1) is an exceptional pattern restricted to specific contexts (e.g., "Ask not what your country can do for you," "Baa baa black sheep," etc.), he or she will raise the main verb across all contexts, albeit with a low probability of 0.01. No English-learning children ever go through a stage

1.2 Where Core Meets Periphery

of main verb inversion: main verb inversion is completely unattested in child English.

Without the ability to recognize exceptions, the learner will have difficulty setting the syntactic parameters. And this is not only a problem for the language acquisition specialist. In recent years, parameters have fallen on hard times because they do not appear as clean and elegant as originally conceived. In a well-known critique of the parameter-based approach to language variation, Newmeyer (2004) considers exceptions an insurmountable challenge. For instance, while French generally places the adjective after the noun (*un livre noir* 'a black book'), there is a special class of adjectives that must appear before it, as shown especially clearly in the contrast between (2b) and (2c):

(2) a. une *nouvelle* maison 'a new house'
     b. une *vieille* amie 'a friend for a long time'
     c. une amie *vieille* 'a friend who is aged'

Additional parameters may be introduced to accommodate the mixed system in French, but the conceptual and the learning problem will not go away. The French facts can be described as one parameter plus lexical exceptions, or two parameters, one for the majority of adjectives and the other for a lexicalized subset, where additional structural patterns and generalizations may be found within (e.g., Cinque 1994). Either way, the French-learning child needs to keep them somehow separate: the general pattern of nominal-adjective order should be established despite the counterexamples in (2b). Similarly, the English-learning child should not allow the occasional and contextually restricted omission of the subject ("mix flour with spices," "had a rough day"; Haegeman 1990) to interfere with the setting of the obligatory-subject parameter. Even more challenging would be a system like Modern Hebrew, which is essentially a null-subject language for first and second person in the past and future tenses but an obligatory-subject language for the rest of the person and tense combinations (Jaeggli and Safir 1989; Shlonsky 2009). Such mixed parametric systems are prima facie evidence against the parameters as overarching "global" properties of languages.

But it is not clear that abandoning parameters is going to help. Suppose one pursues, following Newmeyer 2004, a rule-based approach to syntax. The problem of exceptions remains exactly the same: How does the French learner acquire the *rule*, rather than the *parameter*, that adjectives in general follow the nominal except for those on a finite list (2)? Likewise, it only begs the question to reformulate parameters into a hierarchy of specificity (Baker 2001; Biberauer et al. 2010; Holmberg 2010; Roberts 2012) where some are general, some are construction specific, and still others pertain to individual lexical items. Again, how does a French-learning child know which *vielle*—(2b) vs. (2c)—goes with the restricted parameter and which goes with the general parameter? If pursued to its logical limit, this approach becomes a completely lexicalized theory of language that lists everything (e.g., Sag 2010), which is neither theoretically satisfying nor, as we will see in chapter 2, empirically sufficient.

The alternative route, then, is to salvage the grammar from exceptions. That such a boundary is difficult to draw does not, of course, mean that it does not exist; see Cohn 2006 on the similar conundrum at the juncture between phonetics and phonology. One approach is to reinspect the exceptions: perhaps their idiosyncrasies ought to be assimilated to the core after all (see, e.g., Fodor 2001 for a direct response to some of "nut" cases studied by Culicover 1999). Another approach, one taken here, is to develop a principled demarcation between the core and the periphery, with the recognition that some exceptions are truly accidental and irreducible to general principles (Chomsky and Halle 1968). Such an approach is feasible because children are remarkably unfussed by the core-vs.-periphery problem that has troubled linguists. As will be extensively reviewed in chapter 2 and throughout the book, children are very good at recognizing exceptions at every linguistic level and manage to keep them separate from the core grammar. To wit: Hebrew-learning children can partition the language into two parametric systems from the outset of language acquisition (Levy and Vainikka 2000). For the null-subject component, they behave like children acquiring prototypical *pro*-drop languages such as Italian (Valian 1990). By contrast, for the obligatory-subject portion, they behave like children acquiring prototypical obligatory-subject languages like English (Valian 1990; Wang et al. 1992; Yang 2002), who go through the characteristic stage of subject omission for up to three years.

It is children's remarkable mastery of language that gives hope to theorists: there *must* be a principled division between the grammar and its leaky corners. The current study is a proposal on where the boundary should be drawn.

(3) *Tolerance Principle*
If R is a productive rule applicable to N candidates, then the following relation holds between N and e, the number of exceptions that could but do not follow R:

$$e \leq \theta_N \text{ where } \theta_N := \frac{N}{\ln N}$$

The motivation for the Tolerance Principle is laid out in detail in chapter 3, where I develop a calculus for the price of productivity. The analogy with economics, and hence the title of the current work, is deliberate and I believe appropriate. Just as the price of goods is determined by the balance between supply and demand, I suggest that the price of linguistic productivity arises from the quantitative considerations of rules and exceptions.[1] Drawing extensively from the psycholinguistic literature, I show that exceptions to a rule impose costs to the real-time processing of language. Specifically, learners postulate a productive rule only if it results in a more efficient organization of language, as measured in processing time, rather than listing everything in lexical storage. The Tolerance Principle asserts that for a rule to be productive, the number of exceptions must fall below a critical threshold.

I envision language learning as a search for productive generalizations. Children consider a rule R in their language and evaluate its productivity according to the associated numerical values: N and e, the number of items to which the rule is applicable, and the number of items that defy the rule. The rule is accepted as productive if e is sufficiently small; otherwise learners formulate a revised rule (R') to obtain a new set of values (N' and e') and the Tolerance Principle is applied recursively, as illustrated in Figure 1.1. Thus, the quantitative accumulation of exceptions can lead to the qualitative change in the productivity of rules. Likewise, the core grammar must be able to tolerate a suitably small quantity of exceptions, which will be exiled to the periphery.

1.3 Some Outstanding Problems

In lieu of a roadmap for the materials to come, let me highlight some representative case studies that fall under the purview of the Tolerance Principle.

Just as there are infinitely many sentences, the number of words is also unbounded. Morphology makes compositional use of elemental units, but some processes are clearly open ended while others are severely restrictive. For instance, the English nominalization suffix *-ness* can apply to a broad range of adjectives (*red-redness*), while the suffix *-th* is restricted to only a handful

---

[1] Here I use the term rule to refer to any kind of linguistic generalization that has the potential of open-ended application. In later chapters, I provide explicit formulation of rules, often the output of computational learning models, and quantify their productivity with the Tolerance Principle.

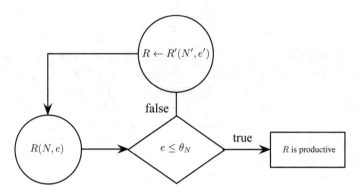

Figure 1.1
The Tolerance Principle as an evaluation measure in language acquisition

of stems (e.g., *warm-warmth*, *wide-width*). Productivity has long been recognized as one of the "central mysteries" in morphology (Aronoff 1976, 35). From the perspective of language acquisition, even innocuous cases become learning puzzles. Take the agentive suffix *-er*, which is unquestionably productive: *teach-teacher*, *drink-drinker*, and more recently, *blog-blogger*. But its productivity can mislead: somehow the child learner must recognize that *rubber* is not for rubbing, *letter* is not someone who lets, and *counter* is (usually) not something that keeps track of numbers. In other words, to learn that *-er* is productive, children must tune out the spurious misapplications embedded in their linguistic experience.

A broad perspective on productivity quickly establishes that a rule's quantitative covergage alone does not guarantee generalization. Consider the English metrical stress system. Thanks to the extensive Latinate vocabulary borrowed into a Germanic language, the stress assignment of English words follows a very complex set of rules (Chomsky and Halle 1968; Halle 1998; Halle and Vergnaud 1987; Hayes 1982). Statistically, however, the English stress system is remarkably simple: about 85% of spoken English words place primary stress on the initial syllable (Cutler and Carter 1987; Legate and Yang 2013). This may tempt learners to postulate a simple default rule that stresses the initial syllable, much like the "quantity insensitive" metrical systems found in languages such as Afrikaans and Chitimacha (Gordon 2002): the residual 15% or so can be lexically stored, resulting in a highly respectable batting average.

But evidently 85% isn't good enough, at least not for long. There is a brief, and very early, developmental stage during which English-learning infants

1.3 Some Outstanding Problems

take the stressed syllable as the beginning of a word (Echols, Crowhurst, and Childers 1997; Johnson and Jusczyk 2001; Jusczyk, Houston, and Newsome 1999) as if they treat the language as invariably stress-initial. However, production studies give unambiguous evidence that by no later than two years and five months (2;5), children are already taking syllable weight into account when assigning stress to words (Kehoe 1997; see Fikkert 1994 for similar findings in Dutch). Furthermore, no theoretical analysis of English stress takes the quantity-insensitive option, which would at least be a great statistical success. Behaviorally, both adults and children take syllable weights as well as lexical categories of words into account to stress novel words (Baker and Smith 1976, Guion et al. 2003, Kelly 1992, Oh, Guion-Anderson, and Redford 2011). It is evident, then, that an overwhelming statistical advantage does not necessarily translate into productivity — so what does?

The transition from one grammar to another, as in the acquisition of English stress, is in fact a very general characterization of child language development. In a celebrated experiment (Berko 1958), young children are shown to use inflectional morphology productively with nonce words (*wug-wugs*, *rick-ricked*). But these results need to be considered alongside findings that morphological rules do not become productive overnight. In an early study, MacWhinney (1975) shows that the development of morphology starts with rote memorization before children discover the productive processes of word formation in their native language. In the acquisition of English, it has been observed that children typically follow a U-shaped learning curve (Marcus et al. 1992; Pinker 1999): irregular verbs are inflected correctly (*hold-held*) early on before succumbing to overregularization (*hold-*holded*) at a later point, which signals the onset of the productive "add -*d*" rule. Apparently children need time and data to accumulate enough regular verbs to counterbalance the irregular exceptions. The trajectory of English past-tense learning is quite typical when considered in the crosslinguistic study of language development (chapter 2). In many (but not all) cases of language acquisition, children show an initial stage of conservatism, during which they do not seem to generalize beyond the input, before the emergence of productivity. How do they calibrate the balance between rules and exceptions? Where is the critical juncture at which children recognize the productivity of rules?

If the *Wug* test puts the unbounded creativity of grammar in the spotlight, then the ineffables in language must be an awkward blemish. In a classic paper, Halle (1973) draws attention to morphological "gaps," the absence of inflected words for no apparent reason. For instance, there are about seventy verbs in

Russian that lack an acceptable first-person singular nonpast form (data from Halle 1973 and Sims 2006).

(4) *lažu 'I climb'
    *pobežu (or *pobeždu) 'I conquer'
    *deržu 'I talk rudely'
    *muču 'I stir up'
    *erunžu 'I behave foolishly'

There is nothing in the phonology or semantics of these words that could plausibly account for their illicit status, yet native speakers regard them as ill-formed. Indeed, defective paradigms such as (4) are hardly rare (Baerman, Corbett, and Brown 2010), even in a morphologically impoverished language such as English—for example, speakers are unsure about the past participle of *stride* (Pinker 1999; Pullum and Wilson 1977). Missing inflections pose considerable challenges for the theories of morphology but a fundamental piece of the puzzle belongs to language acquisition: How do learners know that *some* expected forms are impossible while the combinatorial use of language is in general unimpeded? Where are gaps expected to appear? In other words, upon the presentation of linguistic data, children must deploy a decision procedure that detects productive regularities if present, and comes up empty handed when absent.

Indeed, learning what *not* to say has long been recognized as a critical problem in language acquisition. C. L. Baker, in a well-known study (1979), raises the problem of indeterminate inference in language learning. Of a range of examples he discusses, the English dative alternations have become most prominent:

(5) a. John gave a dish to Sam.
       John gave Sam a dish.
    b. John passed the salami to Fred.
       John passed Fred the salami.
    c. John told a joke to Mary.
       John told Mary a joke.
    d. John donated a painting to the museum.
       *John donated the museum a painting.
    e. *John confessed the police the crime.
       John confessed the crime to the police.

1.3 Some Outstanding Problems

The double-object and prepositional *to*-dative constructions seem interchangeable in the first three examples, but the failure of *donate* and *confess* to do so is unexpected given the semantic similarities of the verbs.

The absence of negative evidence in language acquisition (Brown and Hanlon 1970; Marcus 1993) has led to a considerable body of literature on the acquisition of negative linguistic constraints: How do children know that only some, but not all, unattested forms are ungrammatical? Furthermore, as is reviewed in chapter 6, the distribution of datives across languages (Chung 1998; Levin 2008) and the developmental trajectory of dative acquisition (Conwell and Demuth 2007; Gropen et al. 1989) suggest that these constructions cannot be accounted for solely by innate constraints of syntactic and semantic structures. First, children patiently accumulate evidence about the dative verbs and do not go beyond the adult input (Snyder and Stromswold 1997). Then, very much like the emergence of the "add -*d*" rule, they pounce on a productive rule: in fact, an overly general one that results in errors such as "I said her no" (Bowerman and Croft 2008; Gropen et al. 1989), where *say* was appropriated in the double-object construction. These errors gradually disappear as learners grasp the finer details of the datives. Again, we see a learning process in which the properties of specific items are extended to an entire class: "I texted him an apology" became available as soon as *text* became a verb. At the same time, the grammaticality contrast in (5) suggests that these generalizations must be appropriately constrained.

Finally, let's zoom out from the first few years of language learning by children to the dynamical properties of language across populations and over time. A common (non)response to the problem of exceptions is to appeal to individual variation; as Labov (1972a, 292) laments, "'My dialect' turns out to be characterized by all the sentence types that have been objected to by others." Some exceptional patterns in language may well be a matter of individual and/or dialect variation but they still require a principled explanation. Everyone learns the "-*d*" rule of English past tense as a child, but some learners do so a full year ahead of others (Maratsos 2000; Yang 2002). We cannot understand variability in language acquisition, which has become a major focus in recent research with important practical implications (Hart and Risley 1995), unless we understand how children learn languages in the first place.

Much like the debate over core vs. periphery, there has been a long-standing controversy on the role of rules and exceptions in language change. Is the rise of a linguistic form due to the reorganization of a rule system that systematically applies across the board (Halle 1962; Kiparsky 1965; Kroch 1989; Labov

1994), or does it proceed on an item-by-item and construction-by-construction basis (Bresnan and Ford 2010; Hudson 1997; Wang 1969, 1979)? If there is a division between rules and exceptions, then there must be mechanisms that allow the boundary to blur, because languages are always in a state of variation and change. The very fact of language change entails that children must acquire grammars that are different from those used by their parents. Obviously the difference cannot be wholesale—an English-learning child in Kansas would never acquire the rules of French—how much deviation would be deemed possible, or tolerable, as learners acquire a grammar from the input data generated by another, somewhat different, grammar?

**

These and many other problems form the empirical grounds of the present study. I chose these topics not only because they are amenable to numerical analysis—a necessity for the Tolerance Principle, and the equation in (3) is used almost 100 times—but also because they are traditionally treated with UG-internal solutions. For instance, morphological productivity has been connected to general syntactic principles (e.g., Marantz 2001), the dative constructions are proposed to follow universal syntactic and semantic constraints mediated by innate linking relations (e.g., Pinker 1989), and paradigmatic gaps are produced by shielding specific words from the general rules of language (e.g., Halle 1973). I do not have space to provide detailed assessments of these proposals; rather, I show that they are dispensable and in fact should be dispensed with. The empirical problems are well handled, and indeed unified, by an independently motivated principle of learning: we can do more with less UG, thereby taking a step closer to the solution of both Darwin's and Plato's Problem.

But first, let us understand why the core is worth saving, and why it is ill-advised to focus on exceptions at the expense of rules.

2 The Inevitability of Rules

To a great extent, the current enthusiasm for research on the origin of language is fueled by the perceived linguistic continuity between humans and other species. In a recapitulationist turn, the development of child language is interpreted as retracing the steps of language evolution (Bickerton 1995; Studdert-Kennedy 1998; Wray 1998): Hurford (2011, 590), for instance, regards "language acquisition as the most promising guide to what happened in language evolution." Young children's language has been likened to the signing of primates, because both appear to have limited combinatorial diversity, which has been claimed to follow from rote learning (Terrace et al. 1979; Tomasello 2000a, 2003). According to this view, language emerged when holistic storage shifted to a combinatorial system: studying our children may reveal the evolutionary stages traversed by our ancestors.

These hopes appear misplaced. First, the premise of continuity between children and other great apes is unfounded. When subjected to statistically rigorous analyses, children's language, even in the earliest stage, is in fact combinatorial rather than limited to the storage of specific forms, whereas nonhuman primates never go beyond imitation (Yang 2013a). Second, while no one denies the role of memory in language — at the very minimum, we memorize words — the effect of storage should not be exaggerated. Indeed, were memorization a viable approach to language, Google would have solved all the problems, thanks to its effectively infinite ability to store linguistic data.

The purpose of the present chapter is twofold. First, I review quantitative and computational studies of language to provide a realistic sense of how much, or how little, the storage of linguistic forms can achieve. An abstract and overarching grammar is logically necessary and empirically evident from the very earliest stage of language development. Second, I present crosslinguistic evidence that children occasionally extend productive rules, as is appropriate, but almost never generalize lexicalized forms through analogy. In other words, children are remarkably adept at keeping exceptions from spilling over to the core of grammar; to understand how this is accomplished will require a whole book.

2.1 Statistical Profiles of Grammar

In general, we can only assess the properties of a grammar through the analysis of its behavioral manifestations; the speaker can be queried for grammaticality judgment or put through experimental procedures. But the study of child language faces a unique and more serious challenge: children are not known to be cooperative or patient subjects. Very often, a small sample of children's production is the only, and certainly the most accessible, type of data on hand. But the interpretation of production data is subject to very limiting constraints. Language use is the composite of linguistic and nonlinguistic factors many of which, in a child's case, are still undergoing development and maturation. The moral holds for linguistic study in general: an individual's grammatical capacity may never be fully reflected in his or her verbal behavior.

This much has been well known since Chomsky (1965) drew the competence/performance distinction, possibly even earlier in the Saussurean *langue* vs. *parole*. The pioneers of child language research, including many who did not follow the generative approach, also recognized the gap between what the child knows and what the child says (Bloom 1970; Bowerman 1973; Brown and Bellugi 1964; Brown and Fraser 1963; McNeill 1966; Schlesinger 1971; Slobin 1971; see, especially, the exchanges between linguists and psychologists in the collection edited by Bellugi and Brown (1964)). Two examples from that period illustrate the need to go beyond the observable. Shipley, Smith, and Gleitman (1969) show that children in the so-called telegraphic stage of acquisition nevertheless understand fully formed English sentences better than the reduced forms that resemble their own speech. Brown's synthesis of child language studies available at the time (1973) provides distributional and quantitative evidence against the Pivot Grammar hypothesis (Braine 1963), according to which child language centers around concrete words but not abstract rules — a position that bears more than a passing resemblance to a strand of contemporary thinking. The *item*- or *usage*-based approach to language, most clearly articulated by Tomasello (1992; 2003), holds that children's production data directly reflects their linguistic knowledge, emphasizing the storage of specific linguistic forms and constructions at the expense of general rules and principles.

In this section, I review the statistical properties of language with specific focus on morphology, where the evidence of children's ability to draw appropriate generalizations is especially clear. Some of these properties are very well known in the quantitative study of language, but their significance seems underappreciated in a psychological setting. I show that the available data for language acquisition is sparse: only a small, almost trivial, fraction of possible linguistic forms is available to learners, and children must be able to form wide-ranging generalizations from early on.

2.1.1 The Long Tail

In a perfect world, learning the morphology of a language would be straightforward. Children are presented with full paradigms of word formation, much like those found in linguistic texts or foreign language courses; all they need to do is to figure out the mapping functions among, say, the stem and the inflected forms. For instance, in a language with twelve inflectional forms that involve person, number, and tense, every verb will be fully realized in a complete paradigm table.

In a more realistic setting, however, children are not provided with matched pairs (such as *sing/sang* and *walk/walked*) that specify the forms to be related by morphology.[1] But the more formidable and less studied problem has to do with the sparsity of data. Suppose a child has learned ten verb stems in the hypothetical language with twelve inflectional forms. More often than not, only a tiny fraction of the one hundred twenty possibilities will be available in the linguistic environment. In general, the child will never observe anything close to a complete paradigm: many entries are missing altogether, and the instantiation of the paradigm is in fact the learner's task, rather than the input he or she receives.

The statistics of morphology is very similar to the statistics of words, which are well characterized by Zipf's law (1949); see Yang 2013b for an exposition. Zipf's original study reveals that the frequency of a word is approximately inversely proportional to its rank. Let f be the frequency of a word and r be its rank among N words sorted by frequency. Zipf notices that across all N words, f and r roughly multiply to a constant C. For instance, in the one-million-word Brown Corpus of print materials (Kučera and Francis 1967), the most frequent word is *the*, occurring almost 70,000 times. The second most frequent word is *of*, bearing the rank of 2 and appearing just over 36,000 times. Now 72,000 ($2 \times 36,000$) is not precisely 70,000, but it is not far off. The inverse relationship between rank (r) and frequency (f) is best viewed on a log-log scale. If $r \times f = C$ where C is some constant, then $\log f = \log C - \log r$: the rank and frequency would form a straight line with the slope of -1. Figure 2.1 plots the most frequent 1,000 words in the Brown Corpus on the logarithmic scale.

---

[1] The vast majority of computational models in the study of morphological acquisition, however, require the presentation of paired forms; the task of learning is to identify the mapping by which these forms are related — a very challenging problem in its own right.

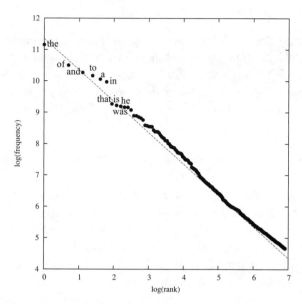

Figure 2.1
Word frequency and rank are inversely related and almost fall on a straight line with the slope −1.0 on the logarithmic scale. Data based on the Brown Corpus, with the top ten words labeled.

Zipf's law has been observed in vocabulary studies across languages and genres, and the log-log slope fit is consistently in the close neighborhood of −1.0 (Baroni 2009). But it is also worth pointing out that while Zipf's law is unquestionably a robust fact about languages, it is not clear how it arises or whether it reveals anything interesting or specific about language, because Zipflike statistical patterns can be observed in other natural and artificial systems (Chomsky 1958; Li 1992; Mandelbrot 1954; Miller 1957). For our purposes, however, Zipf's law has significant implications for the theory of language learning. As can be observed in Figure 2.1, relatively few words are used frequently, indeed *very* frequently, while most words occur rarely, with many occurring only once in even large samples of texts, falling on the long tail. In the Brown Corpus, for instance, 43% of words occur only once, 58% of words occur once or twice, 68% of words occur one to three times. Acquiring a reasonable vocabulary of a language takes a long time, and the learner will have to sit through a lot of boring repetitions.

Language is, of course, not just words; it also has rules that combine words and other units to form meaningful expressions. Here Zipf's long tail grows even longer: the frequencies of combinatorially formed units drop off even

2.1 Statistical Profiles of Grammar

more precipitously. For example, approximately 78% of the Brown Corpus bigrams appear only once and 90% of them appear only once or twice. For trigrams, the singletons and doubles make up 91% and 96% of the unique types respectively. The distribution of words and their combinations in the Brown Corpus is illustrated in Figure 2.2.

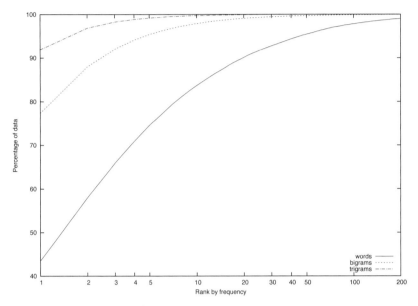

Figure 2.2

The vast majority of n-grams are rare events. The x-axis denotes the frequency of the gram, and the y-axis denotes the cumulative % of the grams that appear at that frequency or lower.

Word formation, which can be viewed as the composition of morphological primitives such as morphemes, shows very similar statistical properties. Figure 2.3 presents the distribution of verbal inflectional morphology in an approximately one-million-word corpus of child-directed Spanish (MacWhinney 2000). The data has been analyzed using a part-of-speech tagger (Freeling; Atserias et al. 2006), which was designed especially for Romance languages. There are 1,584 unique verb lemmas, which altogether appear in 54 inflectional categories.[2] Lemmas are shown in order of increasing rank across the

---

2 The copular and light verbs *estar*, *haber*, *hacer*, *ir*, *ser*, and *tener* are excluded from these counts and from Figure 2.3 because they are generally irregular and do not provide much value

x-axis, inflections are shown in order of decreasing rank across the y-axis, and $\log_{10}$ frequency of specific lemma × inflection combinations is shown along the z-axis.

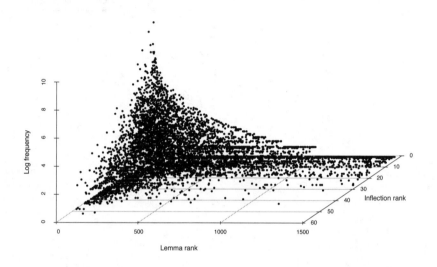

Figure 2.3
Frequencies of CHILDES Spanish lemmas across inflection categories. (Courtesy of Constantine Lignos.)

Several distributional properties clearly emerge from Figure 2.3. First, lemma frequencies broadly show Zipfian characteristics. For instance, the most frequent lemma *ver* almost doubles the number of occurrences of the second most frequent lemma *mirar*, and 521 lemmas — 32.9% of all observed lemmas — only appear once. The top ten lemmas alone account for 42.1% of all occurrences of verbs in the corpus. Second, a similar pattern is visible across inflectional categories; some appear with many lemmas, but more appear with only a few. The most common inflectional category (third-person

---

as exemplar paradigms. An inflectional category is considered a unique combination of person, number, tense, mood, and aspect. Only active-voice verbs are included.

2.1 Statistical Profiles of Grammar

singular present indicative) appears 37,573 times, while two inflectional categories appear only once each: the first- and second-person imperfect subjunctive. Third, looking across the inflections within each lemma, we observe that while the most frequent lemmas appear in many inflectional categories, the least frequent lemmas appear in a few scattered inflectional categories. Visually, if the combination of lemmas and inflectional categories were more uniformly distributed, there would be no trend in the y-axis and few gaps in the plot. Zipf's law seems to pop up everywhere in language.

2.1.2 Quantifying Sparsity

As a means of formalizing data sparsity, consider the metric of inflectional *saturation* introduced by Chan (2008). Saturation is computed per lemma by dividing the total number of inflectional categories observed in the corpus by the number of inflectional categories observed for that lemma. Take, for instance, the verbal system in English. There are six inflectional categories: infinitive, first- and second-person present, third-person singular present, progressive, past tense, and past participle. In virtually all corpora of English, it is easy to find verb lemmas appearing in all of them, providing a complete paradigm and 100% saturation.

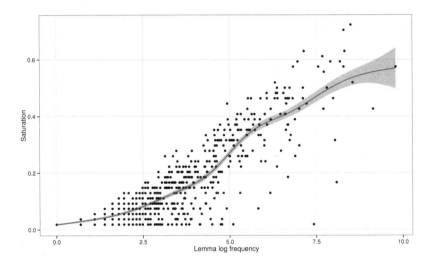

Figure 2.4

Saturation of CHILDES Spanish lemmas across lemma frequencies, with a GAM-derived fit line and standard error estimate. (Courtesy of Constantine Lignos.)

shown in Figure 2.4, in the CHILDES Spanish data under examination, even the most frequent lemmas do not approach 100% saturation. The most saturated lemma is *decir*, with a saturation of 72.2%. The mean saturation across all lemmas is just 7.9%; thus the average verb appears in about one of every thirteen inflectional categories observed in the corpus. A particular combination of lemma and inflectional category is far more likely to be missing than observed.

The phenomenon of low saturation is not specific to Spanish; Table 2.1 presents the analysis of the inflectional morphologies of a dozen languages across genres, including child-directed speech data. As expected, the English corpora contain many verbs with 100% saturation, appearing in all six inflectional categories. For other languages, however, we see that regardless of corpus size, the maximum saturation rate does not even approach 100%; the common case is that the learner does not see a complete paradigm for any verb even with relatively large amounts of input.

2.1.3 Distributional Sparsity and Language Learning

These simple statistical investigations of morphology have direct implications for language acquisition. First, it is unrealistic to expect the full paradigm of any particular stem to be available to child learners (e.g., Albright 2005). Language acquisition takes a finite, and in fact quite modest, amount of input (Hart and Risley 1995, 2003), which, true to Zipf's law, contains relatively few types: learners must generalize fairly aggressively to "fill in" the paradigm table for unattested forms. Second, the sparsity of morphological distribution must be taken into account when we assess the empirical properties of child language.

As we will see in section 2.3, children's inflectional morphology is in general excellent. Their errors tend to be those of omission (e.g., bare stem or infinitive when a tensed form is required) or overuse of a general form (e.g., overregularization of irregular verbs), and the error rates are generally low. But it has been pointed out that low rates of morphological errors do not necessarily imply children's mastery of combinatorial morphology. In an influential line of research, usage-based theorists such as Tomasello (2000a, 2003) note that children's morphology may well result from the storage of lexically specific morphological forms in the input: if the retrieval mechanism is generally reliable, children will also make few morphological errors. In fact, the combinatorial *diversity* in child morphology is quite low, which appears to suggest

2.1 Statistical Profiles of Grammar

Table 2.1
Saturation in a variety of languages. (Adapted from Chan 2008.)

Corpus	Tokens (millions)	Infl. categories	Max. infl. categories per lemma	Max. saturation
Basque	0.6	22	16	72.7
Catalan	1.7	45	33	73.3
Czech	2.0	72	41	56.9
English (Brown Corpus)	1.2	6	6	100.0
English (Wall Street Journal Corpus)	1.3	6	6	100.0
Finnish	2.1	365	147	40.3
Greek	2.8	83	45	54.2
Hebrew	2.5	33	23	69.7
Hungarian	1.2	76	48	63.2
Italian	1.4	55	47	85.5
Slovene	2.4	32	24	75.0
Spanish	2.6	51	34	66.7
Swedish	1.0	21	14	66.7
CHILDES Catalan	0.3	39	27	69.2
CHILDES Italian	0.3	49	31	63.3
CHILDES Spanish	1.4	55	46	83.6

the absence of productive rules but only the access of lexically specific morphological forms. For instance, Pizzuto and Caselli (1994) find that in a corpus of child Italian speech, 47% of all verbs used by three young children (1;6 to 3;0) are used in one person-number agreement form, and an additional 40% are used with two or three forms, where six forms are possible (3 person × 2 number). Only 13% of all verbs appear in four or more forms. The low level of combinatorial diversity has been taken as a major source of evidence for the usage-based theory of language learning (Tomasello 2000a).

Here the sparsity of child morphology must be viewed in the context of morphological sparsity in general. Table 2.2 summarizes the results from the corpus analysis of the child and child-directed speech in Italian, Spanish, and Catalan currently available in the CHILDES database (MacWhinney 2000). The morphological data is again analyzed with the Freeling tagger, and only

Table 2.2
Verb agreement distributions in child and adult Italian, Spanish, and Catalan

Subjects	1 form	2 forms	3 forms	4 forms	5 forms	6 forms	S/N
Italian children	81.8%	7.7%	4.0%	2.5%	1.%7	0.3%	1.5
Italian adults	63.9%	11.0%	7.3%	5.5%	3.6%	2.3%	2.5
Spanish children	80.1%	5.8%	3.9%	3.2%	3.0%	1.9%	2.2
Spanish adults	76.6%	5.8%	4.6%	3.6%	3.3%	3.2%	2.6
Catalan children	69.2%	8.1%	7.6%	4.6%	3.8%	2.0%	2.1
Catalan adults	72.5%	7.0%	3.9%	4.6%	4.9%	3.3%	2.3

tensed forms are counted. Each cell represents the percentage of verb stems that are used in one, two, three, four, five, and six person and number forms.

As can be observed in Table 2.2, Spanish and Catalan children and adults show very similar, and very low, usage diversity in their agreement morphology. Italian children use more stems in only one form than Italian adults (81.8% vs. 63.9%), but this can be attributed to the measure in the last column, which is the ratio between the total number of inflected forms (S) over the total number of stems (N) — that is, the token/type ratio. This value represents the average number of opportunities for a stem to be inflected: when the ratios are comparable between children and adults, as in the Spanish and Catalan data, we observe similar diversities of agreement morphology. In the case of Italian, the adults had roughly two-thirds more opportunities to use a stem than the children; it is not surprising that adults produced fewer one-form verbs than children. It is premature, then, to conclude that low diversity implicates an underdeveloped morphological system: *all* morphological systems are sparsely represented. Of course, the quantitative similarity between child and adult morphologies does not conclusively show that children's grasp of agreement morphology is identical to that of adults. To do so, one must compare the statistical expectations of rule-based morphology against the usage profile of children's morphology to show statistical agreement; I review a case study of syntactic acquisition in chapter 7.

Research in computational linguistics also lends support to the indispensability of abstract rules in language. Statistical models of language can provide a useful way to assess how different conceptions of linguistic structures contribute to empirical coverage. For instance, a statistical model of grammar, such as a probabilistic parser (Charniak and Johnson 2005; Collins 1999), encodes several types of grammatical rules: a phrase *drink water* may be represented in multiple forms ranging from categorical (VP → V NP) to lexically specific (VP → V_{drink} NP) or bilexically specific (VP → V_{drink} NP_{water}). These multiple representations can be selected and combined to test their descriptive effectiveness. It turns out that most generalizing power comes from categorical rules; lexicalization plays an important role in resolving syntactic ambiguities (Collins 1999) but bilexical rules — lexically specific combinations that form the cornerstone of usage-based theories (e.g., Tomasello 1992; see also Sag 2010) — offer virtually no additional coverage (Bikel 2004). The broad coverage provided by abstract rules, and the marginal improvement due to lexically specific combinations, can be easily observed in Figure 2.5.

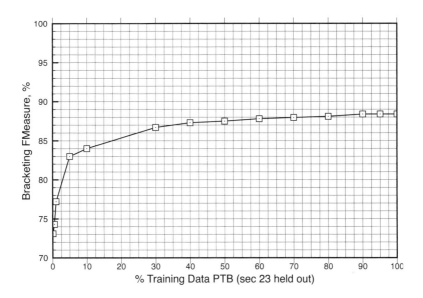

Figure 2.5
The quality of statistical parsing as a function of the amount of training data. (Courtesy of Robert Berwick and Sandiway Fong.)

In Figure 2.5, a statistical parser (Bikel 2004; Collins 1999) is trained on an increasing amount of data (x-axis) from the *Wall Street Journal* portion of the Penn Treebank (Marcus et al. 1999), and its performance (y-axis) is measured on a portion of the data held out for testing. More data does improve parsing quality, but the most striking feature here is found toward the lower end of the *x*-axis. The vast majority of data coverage is gained on a very small amount of data, between 5% and 10%, which can *only* be attributed to highly abstract and general rules because the parser will have seen very few lexically specific combinations. Thus lessons from the engineering application of language are strongly convergent with the conclusions from our exploration of child-directed language: memorization of specific linguistic forms is of very limited use in both morphological and syntactic learning and cannot substitute for the overarching power of a productive grammar.

It is not surprising, then, to find that children are excellent rule learners: they must be, for otherwise language acquisition would not be possible at all, never mind in a few short years. To understand how rules are learned, let's first review how well children learn them across languages.

2.2 Interlude: Irregular Rules and Irregular Verbs

No review of morphological acquisition would be complete without a discussion of English past tense, one of the most extensively studied topics in all of linguistics and cognitive science (Pinker 1999). The impoverished morphology of English inflections hardly seems appropriate to appreciate the complexity of languages, but as we will see, the study of English past tense has raised central questions for the general theory of language and language acquisition. The treatment here, however, will be relatively brief. My main purpose is to establish the plausibility that irregular past tense is not based on word-based association according to the prevailing view in psycholinguistics, but is formed by rule-based computation. This is a very traditional conception of irregular morphology that will have important consequences in our discussion of productivity and the motivation of the Tolerance Principle.

In the traditional linguistics literature that can be traced back to the writings of Sweet (1892), Jespersen (1942), and Hockett (1942), the past tense of English is treated as a computational system that takes the stem (e.g., *walk* and *think*) as input and produces the past tense (e.g., *walked* and *thought*) as output. Bloch 1947, for instance, divides the verbs into inflectional categories: the stem *think* falls into a category that adds the -*t* suffix (his category B3, p. 413) and follows a morphophonemic alternation (his Type II, p. 415) that changes the vowel to /ɔ/. According to this view, both regulars and irregulars are generated by an input-output computational system; it is just that the latter requires special instructions because their categories and alternations are not predictable. Generative grammar has largely preserved this tradition from *The Sound Pattern of English* (Chomsky and Halle 1968) to Distributed Morphology (Halle and Marantz 1993). The more recent formulations tend to be more abstract than Bloch's treatment, which generally pertains to the surface pattern. In the present discussion, the term rule simply refers to a generative procedure: some rules are productive and apply to an open-ended set of items ("add -*d* to form past tense"), while others are lexically restricted to a finite list, which will be referred to as *irregular* rules (cf. morpholexical in the sense of Anderson 1974 and Lieber 1981). For instance, Bloch's category B3 and Type II form an irregular rule, one that does not apply beyond *bring*, *buy*, *catch*, *fight*, *seek*, *teach*, and *think*.

The traditional, completely rule-based, approach to English verbal inflection did not feature prominently in the past tense debate (Clahsen 1999; Marcus et al. 1992; McClelland and Patterson 2002; Pinker and Prince 1988; Pinker and Ullman 2002; Rumelhart and McClelland 1986) until the early 2000s (Albright and Hayes 2003; Yang 2002). These studies agree with the dual-route model (e.g., Clahsen 1999; Pinker 1999) in upholding the reality of the regular rule, but differ from it (and the connectionist approach) in the treatment of the irregulars.[3] Here the two sides of the past tense debate are in agreement that irregular past tense is learned as paired associations between the stem and the past form (e.g., *think∼thought*, *hold∼held*). The rule-based approach regards all past tense as rule generated; the difference between regular and irregular verbs is that the regulars are computed by a productive rule and the irregulars are computed by a set of unproductive irregular rules. I briefly review the evidence in favor of the rule-based approach to irregular verbs with two goals in mind. First, establishing the reality of irregular rules is crucial for understanding the price of productivity, as I discuss in detail in chapter 3. Second, I believe that linguistic theories have much to offer to the psychological study of language, but this can only be done by engaging seriously with the materials outside theoretical linguistics.

What evidence would bear on the nature of irregular past-tense learning — by word association or rules? The strongest evidence for associative storage comes from frequency effects. In general, irregular past-tense forms that are

---

3 I single out the dual-route model because it is the most visibly engaged in the past tense. It was preceded by proposals, including those of Jackendoff 1975, Kuczaj 1977, Bybee and Slobin 1982, and to some extent Anshen and Aronoff 1988, that make similar claims regarding the irrgular verbs.

more frequently used in the input tend to be acquired more accurately by children.

Logically, children may make three kinds of errors regarding the use of the past tense.[4]

(1) a. Overregularization: an irregular verb is inflected with the "-*d*" rule (e.g., *hold-holded*).
    b. Overirregularization
        i. Regulars: a regular verb is inflected with an irregular rule (e.g., *heal-helt* along the lines of *feel-felt*, *wipe-wope* along the lines of *ride-rode*).
        ii. Irregulars: an irregular verb is inflected with a nontarget irregular rule (e.g., *sting-stang* along the lines of *sing-sang*, *write-writ* along the lines of *bite-bit*).

Overirregularization errors, which are sometimes referred to as analogical errors, are frequently alluded to but rarely studied systematically. They turn out to be virtually absent; I return to this matter in section 2.3 when I discuss the role of productivity and analogy in child morphology. Overregularization errors, by contrast, are among the best-documented phenomena in the study of language acquisition. Earlier studies suggest that 4.2% of irregular verbs children produce are over-regularized (Marcus et al. 1992; Pinker 1995), but later studies have found somewhat higher rates. For instance, Yang (2002) reports a rate of 10% based on approximately 7,000 tokens from four large longitudinal corpora in the public domain, and Maslen et al. (2004) find that 7.8% out of about 1,300 tokens produced from a single child are overregularized. In general, it has been found that irregular verbs that appear in the past tense more frequently in adult speech tend to have lower rates of overregularization. Marcus et al. (1992) report a negative correlation of -0.33 between an irregular verb's frequency in adult language and its error rate in children's production.

However, a correlation of -0.33 is hardly a compelling statistical result. Partly motivated by this weakness, Yang (2002) explores the plausibility of the traditional rule-based approach to irregular past tense. The learning of irregular verbs consists of several components:

---

4 By errors, I mean forms that deviate from the "standard" adult usage form: unfortunately, the acquisition data presently available does not offer any opportunity to systematically study the acquisition of dialectal variation in past-tense formation (Anderwald 2009).

(2) a. The learner must construct rules for the irregular verbs such as those found in Bloch 1947.
   b. The learner must associate each irregular stem with the corresponding irregular rule.
   c. To inflect an irregular past tense, the learner needs to locate the associated irregular rule and then apply it.
   d. The failure to locate or apply the irregular rule results in the use of the default rule of "add -*d*" and thus overregularization.

The rule-based approach does not dispense with the role of memory: the irregular verbs are unpredictable and their association with the corresponding irregular rules can be formed through some form of rote learning. However, the rule-based approach makes significantly different predictions from the word-based approach. For instance, the learner receives evidence for the "ought" rule, which applies to *bring, buy, catch, fight, seek, teach,* and *think,* whenever any of the seven verbs appears in the past tense. This leads to the notion of a *rule frequency,* which is the sum of the frequencies of the inflected forms; for additional discussion, see section 3.3.2.1. The word-rule association (i.e., the lexicalization aspect of irregular rules) can only be established on exposure to the specific verbs: *brought* will contribute to the lexicalization of *bring,* and only *bring,* although it simultaneously adds to the weight of evidence for the "ought" rule, which is shared by other members of the class (*think, catch,* etc.).

The bipartite nature of the rule-based approach contrasts with the word-based associative approach. Consider two irregulars that are comparable in frequency but where one belongs to a very frequent irregular rule and the other belongs to a relatively infrequent irregular rule. For the word-based theory, the two verbs are expected to be learned at a comparable level of accuracy. For the rule-based theory, the verb belonging to the more frequent rule should be learned better because the rule, being collectively and more abundantly attested in the input, will be used more reliably, as indicated by step (2c) in the computation of irregular verbs (2). The irregular verbs *taught* and *flew* are two such examples. They are comparable in frequency (81 and 117 in 3.6 million words of English spoken by the mothers in the CHILDES database), but children overregularized *flied* 26% of the time (23/89) but *teached* only 12% of the time (3/25). (Many additional examples can be found in Yang 2002.) This performance disparity is hard to explain under the word-based theory,[5] but is straightforward under the rule-based theory. The verb *teach* belongs to the "ought" rule which includes very frequent members such as *thought, brought, bought*, and *caught* (with frequencies of 1516, 416, 367, and 288 in the same corpus); by contrast, the most frequent member of the rule that includes *fly-flew* is *know-knew*, which only appears 390 times. This sort of free-rider effect goes a long way toward ameliorating the weak statistical correlation between input frequencies and overregularization rates.

In earlier work, I have identified another pattern that further suggests that past-tense acquisition is not a matter of building word associations (Yang 2002). A subset of English irregular verbs has long been analyzed as following more general morphophonological processes. Specifically, verbs such as *hide-hid, feel-felt*, and *say-said* have been treated as instances of the English vowel-shortening process triggered by suffixation (null, *-t*, and *-d* respectively), similar to alternations found in *deep-depth, nation-national, divine-divinity*, etc. (Halle and Mohanan 1985; Myers 1987). Intriguingly, all vowel-shortening verbs, including those with very low input frequencies, are used very well by children, with an overall overregularization rate of only 2%, compared to 10% across all verbs (Yang 2002). The learning mechanism and the resulting representations for the vowel-shortening irregulars are by no means well understood, but it is plausible that vowel shortening, being a fairly general process in English morphophonology, can draw support from data beyond the realm of the past tense, which in turn benefits the acquisition of those irregular verbs that make use of the same process.

These preliminary investigations are only beginning to unravel the linguistic complexity in a very small corner of English morphology. But they seem to have shown that linguistic theories may directly, perhaps even isomorphically, bear on the psychological reality of language learning and use. The rule-governed nature of irregular verbs reviewed here goes hand in hand with recent behavioral and neurological evidence for the compositional formation of irregulars (see, e.g., Allen and Badecker 2002; Fruchter, Stockall, and Marantz 2013; Morris and Stockall 2012; Regel et al. 2015; Stockall and Marantz

---

5 Especially because the members of *taught*'s class are completely heterogeneous while the words patterning with *fly-flew* have some phonological similarity (e.g., containing a long vowel as in *blow-blew*, etc.), which should mutually facilitate learning under an associative/analogical account.

2006), as researchers pursue a more theoretically informed approach to psycholinguistics.

This brief review of English irregular verbs under the rule-based approach raises some important and unsettled questions. How are rules — both regular and irregular — inductively constructed by child learners? Some of the existing work is reviewed in chapter 3 but much remains unclear. Additionally, how do learners know that the irregular rules are lexically limited and do not extend to novel items while the "add -*d*" rule is productive and open ended? Notice that this issue also needs to be resolved by the dual-route model (Pinker 1999): without knowing what the default rule is, learners do not know whether to commit a verb to associative storage or to apply a rule without needing to commit any additional memory. As I discuss below, these questions are central to morphology and language acquisition.

2.3 Productivity in Child Language

The acquisition of morphological productivity can be summarized very succinctly: children draw a sharp, essentially categorical, distinction between productive and unproductive processes.

2.3.1 The Wug Test

The simplest assessment of productivity is the celebrated Wug test. In a landmark study, Berko (1958) introduced young children to a wide range of novel words including nouns, verbs, and other categories and recorded their responses in the following elicitation task.

(3) This is a WUG.
 Here is another one.
 These are two _____.
 WUGS.

The Wug test is now widely used (e.g., Albright and Hayes 2003; Bybee and Moder 1983; Bybee and Slobin 1982; Clahsen 1999; Hahn and Nakisa 2000; Hayes et al. 2009; Marcus et al. 1992; Becker, Ketrez, and Nevins 2011), which has had considerable influence in linguistic theorizing (Hay and Baayen 2005; Taylor 2003; Tomasello 2003). But it is unclear to what extent this test provides a direct window into language and its acquisition. First, children are far from perfect in their Wug performance (Berko 1958, 160). For instance, in the regular past inflection of novel verbs, first graders successfully produced *rick-ricked* and *spow-spowed* in as few as 25% of the test cases. While this appears to be a sign of morphological deficiency, it is important to remind ourselves that most if not all English-learning children acquire the "add -*d*" rule by the age of three (Marcus et al. 1992), as indicated by their spontaneous use of overregularized forms (e.g., *fall-falled*). There seems to be a gap of at least three years between acquiring the regular past tense and using it in a specific experimental design: failing to pass the Wug test does not imply imperfect knowledge of morphology.

When children do produce a response in the Wug test, however, their behavior points to a categorical distinction between regular/productive and irregular/unproductive processes. For instance, when children failed to produce *spowed* for *spow*, they produced no response at all rather than, say, *spew*, which would follow the analogy of *know-knew*. It is now largely forgotten that Berko also systematically investigated the role of analogy. Children were presented with novel verbs such as *bing* and *gling* that are strongly similar to existing irregular verbs (*sing-sang*, *sting-stung*, etc.), which would have great potential for analogical extension.

(4) This is a man who knows how to GLING.
He's GLINGING.
(Picture of a man exercising.)
He did the same thing yesterday.
What do he do yesterday?
Yesterday he _____.

Children were overwhelmingly conservative in their responses. Only one of the eighty-six children in Berko's study supplied the analogical form *bang* and *glang*. Berko did notice that adult subjects were far more likely to extend the irregular form. But as discussed at length by Schütze (2005) and others, this finding may well be an effect of the task rather than evidence that adults have productive irregular rules. When adults acquire English as a second language, overregularization is also the dominant error patterns and overirregularization is almost never documented (e.g., Pica 1983). This is line with findings in first language acquisition by children, as I discuss further in section 2.3.2.

Despite its simplicity, then, the Wug test can only be regarded as an indirect reflection of morphological knowledge, filtered through an ill-understood experimental design and other performance factors. Fortunately, the abundance of naturalistic child language data sidesteps these methodological complications and provides unambiguous evidence for how children deal with morphological productivity. As usual, our discussion starts with the much-studied English past tense.

2.3.2 Regularization vs. Irregularization in English

There is no doubt that the "*-d*" rule in English past tense is productive: when new words such as *google* were introduced into the language, the regularly inflected past tense (*googled*) became instantly available.

Readers may have the impression that irregular rules are also frequently extended. The apparently recent emergence of *snuck* as the past tense of *sneak*, along the lines of *strike-struck* and *stick-stuck*, is an often-cited example (Bybee and Moder 1983), but as we will see in chapter 5, its historical development does not provide unequivocal evidence for analogical change (Anderwald 2013). The language acquisition evidence, however, is completely unambiguous. Children almost never generalize the irregular patterns beyond the appropriate lexicalized list. The frequent reference to analogical errors such as *bite-bote*, *wipe-wope*, *think-thunk*, etc. (Ambridge et al. 2015; Bowerman 1982; Bybee 1985; Pinker 1999; Pinker and Prince 1988) is almost completely anecdotal. As a matter of fact, the empirical evidence for overirregularization is very slim: there is not a single attested example of these in the entire CHILDES database of almost two million words of child English. The most comprehensive empirical study of analogical errors (Xu and Pinker 1995) in fact refers to these as "weird past tense errors" on the basis of their rarity. Xu and Pinker examined over 20,000 past-tense tokens produced by nine children, and only forty weird errors (0.02%) were identified, which is at least an order of magnitude lower than the rate of overregularization; see section 2.2. The near total absence of overirregularization has received surprisingly little attention in the past-tense literature. For instance, computational modeling, which kicked off the past-tense debate (Rumelhart and McClelland 1986), is devoted almost entirely to the phenomenon of overregularization; all connectionist networks examined by Marcus (1995) over-irregularize at a level much higher than children. Similarly, a computational study by O'Donnell (2015) tests several nonconnectionist models of past-tense learning. The models are presented with both existing and novel words in the stem form and generate an inflected past tense form as output. Most models are reasonably successful at passing the Wug test for regular verbs, but they all produce a high number of analogical forms on the basis of the existing irregulars. The best model overall, a Bayesian model (O'Donnell 2015) that tries to strike a balance between lexical storage and rule-based computation (Johnson, Griffiths, and Goldwater 2006), produces 10% of overirregularization patterns on novel items, two orders of magnitude higher than the overirregularization rate by human children (Xu and Pinker 1995). In fact, the model's overirregularization rate is comparable to English-learning children's overregularization rate, completely missing a key result from past empirical research.

A closer examination of the irregularization errors, exhaustively listed in the Xu and Pinker study, suggests an even lower rate. Of the forty attested examples, at least ten are very likely speech errors (e.g., *fit-feet, say-set, fight-fooed, bite-bet*). Furthermore, examples such as *sleep-slep* are counted as incorrect application of an irregular rule but they are more likely due to the word-final t/d-deletion process in spoken English (Labov 1972b, 1989). The only systematic error follows the *ing→a/ung* pattern, and the verb *bring* as *brang* or *brung* is the only item that is frequently overirregularized. It is possible that children form a productive rule that changes /ɪ/ to /æ/ before /ŋ/: for children's small vocabulary, this pattern is consistent with two out of the three verbs that fit the structural description (*ring, sing* and *bring* ; see section 4.1 for detailed discussion). But it is also possible that *brang* is present in the input as a matter of dialect variation (Herman and Herman 2014): in fact, *brang* can be found in the child-directed speech (MacWhinney 2000). In any case, the drastically different rates of overregularization and overirregularization suggest that there is a (near) categorical distinction with respect to productivity between productive rules and irregular rules.

2.3.3 Productivity across Languages

The productivity/analogy asymmetry has been observed in numerous studies of child language, mostly in the arena of morphology but similar findings have been reported for syntax as well.[6] When children make mistakes, they almost always employ a default or productive form (e.g., *thinked*) or omit the appropriate form altogether: they almost never substitute an inappropriate form. Of course, the very notion of a default raises complex theoretical questions and

---

6 Recall the restrictions on null subjects and verb movement in child English discussed in section 1.2; the learner somehow recognizes exceptions as such, which never spill into the terrain of rules.

2.3 Productivity in Child Language

I will in due time review the acquisition of what appears to be defective, or defaultless, morphological systems. But first, an overview of previous findings on productivity from the crosslinguistic research on child morphology.

In a study that targets the German agreement affixes *-st* (2SG) and *-t* (3SG), Clahsen and Penke (1992) find that while a child ("Simone") supplies an agreement affix in obligatory context only 83% of the time, almost all the errors are those of omission. When the child does produce an agreement affix, it is almost always the appropriate one (over 98% of the time); inappropriate use (e.g., substituting a *-t* for *-st*) is virtually absent. Similar patterns can be observed in the acquisition of Italian. In a cross-sectional study (Caprin and Guasti 2009, 31), children in all age groups use a diverse and consistent range of tensed forms. Furthermore, the use of person and number agreement is essentially error free throughout, reaching an overall correct percentage of 97.5%, consistent with previous reports (Guasti 1993; Pizzuto and Caselli 1994). Children's impressive command of agreement is most clearly seen in the acquisition of languages with considerable morphological complexities. In a study of morphosyntactic acquisition in Xhosa (Gxilishe et al. 2007), children are found to gradually expand the use of subject agreement across both verbs and noun classes. The rate of marking in obligatory contexts as well as the diversity of the morphological contexts themselves steadily increased. In a process best described as probabilistic, the children often alternate between marking a verb root in one instance and leaving it bare in another, very much like the use/omission alternation pattern reviewed earlier. Crucially, however, virtually all agreement errors are those of omission: 139 out of 143 or 97.2% to be precise. Substitution errors are again very rare, confirming previous research on languages with similarly complex morphology (Deen 2005; Demuth 2003), including polysynthetic languages such as Inuktitut (Allen 1996).

I now turn to several case studies that focus more specifically on the contrast between regular and irregular morphologies in children's naturalistic speech. This type of evidence has been accumulating in the literature on the dual-route approach to morphology (Clahsen 1999; Pinker 1999), for which a categorical distinction between regular and irregular processes is of central importance. The evidence is unequivocal.

The German participle system consists of a productive default *-t* suffix (*fragen-gefragt* 'ask-asked'), as well as an unpredictable set of irregulars taking *-n* (*stehlen-gestohlen* 'steal-stolen') (Wiese 1996); more on this in section 3.3.2.2 and chapter 4. In a series of studies, Clahsen and colleagues (Clahsen

1999; Clahsen and Rothweiler 1993; Weyerts and Clahsen 1994) find that children across all age groups overapply the *-t* suffix to the irregulars, where the reverse usage is virtually absent. Their longitudinal data contains 116 incorrect participle endings, out of which 108 are *-t* errors (**gekommt* instead of *gekommen* 'come', i.e., overregularization). The rest are irregularization errors such as **geschneien* for *geschneit* (snowed). According to the authors, the overall rate of *-t* regularization is 10% of all usage, which suggests that the *-n* irregularization rate is merely 0.75% (based on the 8 *-n* errors compared to 108 *-t* errors). The acquisition of German past participles, therefore, is quite analogous to that of English past tense reviewed earlier (Xu and Pinker 1995), because both point to the productive asymmetry between regulars and irregulars.

The inflection of Spanish verbs provides a complex but highly informative case for exploring productivity in child language; see section 5.1.2 for additional discussion. In Spanish, stems generally consist of theme vowels and roots, which are then combined with affixes for inflection. For instance, a finite form of the verb *hablar* (to talk) is *habl-á-ba-ais*, which represents the root (*habl* 'speak'), the theme vowel (*a*), the past tense (*ba*) and the second-person plural (*ais*). The theme vowels define three conjugations, with the first (*a*) being the numerically dominant class, followed by the second and third (Real Academia Española 1992, cited in Clahsen, Aveledo, and Roca 2002). The irregularity in Spanish inflection comes in two broad classes concerning the stem and the suffix respectively. There are some thirty verbs that are highly irregular with the insertion of a velar stop in certain inflections. These examples include *tener* (to have), *poner* (put) or *salir* (go out), whose first-person singular forms are *tengo*, *pongo*, and *salgo*, respectively. The majority of irregulars undergo a well-known morphophonemic alternation known as diphthongization, a process which is not limited to verbal morphology (Eddington 1996; Harris 1969). For these verbs, the mid-vowel is diphthongized in stressed syllables. (5) shows the pattern for the present indicative of the verbs *comenzar* (to begin) and *contar* (to count), where the acute accent marks stress.

(5) comiénzo comiénzas comiénza comenzámos comenzáis comiénzan
     cuénto   cuéntas   cuénta   contámos   contáis   cuéntan

While it has been suggested that the *form* of diphthongization is predictable (Harris 1969), the verbs that undergo diphthongization are not predictable and must be lexically learned. It is in fact possible to find minimal pairs such as *contar-montar* where the former contains the diphthong (*cuento*) but the

2.3 Productivity in Child Language

latter does not (*monto*). And there are a few common verbs that show both diphthongization and velar insertion in some forms. For instance, *tener* (to have) and *venir* (to come) show velar insertion in the present subjunctive and first-person singular of the present indicative and [ie] diphthongization in the second-person singular and the third person of the present indicative. Although inflectional irregularities in Spanish mostly concern the stem, the suffixes are affected as well. For the stem *querer* 'to want', for instance, the 1SG past tense is *quise*, which involves the stem change noted earlier but also takes an irregular suffix rather than the regular suffix, which would have resulted in **quisí*. The suffix in the 3SG past tense *puso* 'she/he/it put' is -*o* and the regular suffix would have formed **pusió*.

Clahsen et al. (2002) analyzed the verbal inflections of the fifteen Spanish-speaking children and found strong evidence for a categorical distinction between the regular and irregular inflections.

(6) a. The irregulars: children produced a total of 3,614 irregular verb tokens, out of which 168 (4.6%) are incorrect either in stem formation or suffixation.
  i. Of the 120 stem-formation errors (see below), 116 are overregularizations and only one is analogical irregularization.
  ii. Of the 133 suffixation errors, 132 are overregularizations with no occurrence of irregularization.
 b. The regulars: children produced 2,073 regular verb tokens, only 2 of which are the inappropriate use of irregular suffixes.

Collectively, then, the rate of analogical irregularization is only 0.001% for all verbs, and also 0.001% for the irregulars: again, orders of magnitude lower than the rate of overregularization errors.

Clahsen et al.'s study does not include errors regarding diphthongs; all the stem-formation errors are failures to use a diphthong when required. Although this broadly supports the notion of diphthongization as a lexicalized process, it does not consider the possibility of "mis"-diphthongization — for example, the child produces [ie] alternation when the correct diphthong is [ue]. To address this issue, Mayol (2007) provides a finer-grained investigation of inflectional errors focusing more specifically on the types of stem-formation errors and their underlying causes. The speech transcripts of six Spanish-learning children, almost 2,000 tokens in all, fail to yield a single misuse of diphthongization.

Finally, I would be remiss if I failed to mention some perennial challenges in the study of morphology. As much as children quickly and accurately grasp the productive aspects of language, they also run aground in the corners of the grammar where productivity unexpectedly fails. Just as the existence of paradigmatic gaps challenges theories of morphology, the issue of nonproductivity has also turned up in the acquisition literature. In an important study, Dąbrowska (2001) shows that in Polish, masculine nouns in the genitive singular either take an *-a* or *-u* suffix, but neither is the default according to the standard tests for productivity. Moreover, children make very few errors with either suffix, a pattern consistent with the acquisition of unproductive (i.e., lexicalized) morphological processes reviewed earlier. The Polish case is of considerable theoretical interest. It suggests that child learners should not presuppose the existence of a default rule as suggested by the dual-route model of morphology (Clahsen 1999; Pinker 1999). The absence of a default also poses challenges to competition-based theoretical frameworks such as Distributed Morphology (Halle and Marantz 1993) and Optimality Theory (Prince and Smolensky 2004) in which a winning form is always expected to emerge. I address the acquisition of Polish genitives and the general problem of defective morphology in chapter 5.

In summary: the distinction between productive and unproductive aspects of morphology appears completely categorical in child language development. This conclusion contrasts with claims about the probabilistic and gradient nature of productivity (Hay and Baayen 2005). In the words of McClelland and Bybee (2007, 439), "there is no dichotomous distinction between productive and unproductive phenomena; rather, there are only degrees of productivity." How does the gradient view square with the empirical findings of child language, which show near total categoricity? In my view, the root of these contrasting conclusions about productivity is methodological. In some cases, the metric with which productivity is measured *guarantees* gradient results. For instance, in a long line of work starting with Baayen 1989, productivity is quantified as a ratio between two numerical values of lexical statistics. Putting aside the empirical suitability of this approach (see section 3.5.4 for discussion), productivity measured as such necessarily falls between the values of

2.3 Productivity in Child Language

0 and 1, precluding the categorical view of productivity.[7] Similarly, behavioral measures of productivity such as Likert-scale rating for inflected forms (Albright and Hayes 2003; Hahn and Nakisa 2000; Prasada and Pinker 1993; Ramscar 2002) are also biased toward a gradient interpretation. But experimental psychologists have long known that categorical tasks are likely to elicit categorical responses, and gradient tasks such as rating are likely to elicit, alas, gradient results. Participants in rating tasks are inclined to spread responses over whatever range they are given (Parducci and Perrett 1971). In a classic study, Armstrong, Gleitman, and Gleitman (1983) find that gradient judgment can be obtained even for uncontroversially categorical concepts such as "even number." Thus, gradient results in the investigation of morphological productivity do not mean that productivity itself is in fact gradient (see Schütze 2005, 2011 for additional discussion). Note that I do not reject gradient tasks as appropriate tools for studying certain aspects of linguistic knowlege and use. For instance, speakers may have intuitions about the relative frequencies of linguistic forms — a gradient fact — that may be reflected in behavioral measures (e.g., Trueswell 1996). But the very method, which guarantees a gradient outcome, cannot be used to determine whether a phenomenon such as productivity is gradient or categorical.

In light of the results from child language, and following scholars such as Botha (1969), Bauer (1983), and others, I propose to do away with notions such as "semiproductivity" that one occasionally finds in the theoretical and psycholinguistic literature (Jackendoff 1996; Matthews 1974; Pinker 1991). If a semiproductive rule is one that has exceptions — as opposed to those that are exceptionless — then virtually all rules are semiproductive, including those regarded by all as fully productive (such as the "-*d*" rule in English), and the notion of semiproductivity becomes vacuous. If the term refers to rules whose productivity status is gradient or fuzzy, then it is immediately contradicted by the crosslinguistic acquisition evidence that productivity, at least in child language acquisition, is categorical, without fuzzy boundaries: the "continuum" alluded to by McClelland and Bybee (2007) does not seem to exist. A final

---

[7] It is in principle possible that there is a threshold $0 \leq \varepsilon \leq 1$ such that a productivity ratio above ε is categorically productive and below ε categorically unproductive (i.e., lexicalized). I do not believe this likely. As noted in chapter 1, an overwhelming coverage of words, such as the rule that places the primary stress on the initial syllable of English words, may fail to be productive. By contrast, a rule that applies to a very small proportion of words, such as the -*s* suffix in German noun plurals, does achieve productivity (Marcus et al. 1995), a case I study in detail in chapter 4. Thus, there appears to be no one-size-fits-all threshold to separate productive from unproductive processes.

possibility is that semiproductivity refers to individual variation regarding the productivity of a rule, which would also contribute to the impression of gradience, especially if individual results are pooled and averaged (as is frequently the case in experimental research). This is a genuine possibility, and as we will see in section 4.1.2, different English-learning children appear to acquire the productivity of the rules at quite different stages of development. From the perspective of the individual grammar, however, there is no point at which a rule is, say, 75% productive; it is difficult to see what such a notion would even mean.

****

As I conclude my defense of the combinatorial nature of language, the need for a learning model becomes even greater. Words and rules do not come prepackaged: "This is an irregular verb," "That's the default rule." Children need to sort these out on their own. Sometimes a child needs to fail to succeed, as in the curious case of gaps just reviewed, where no alternation should emerge as the productive default. And the model must get it right quickly: the child only has a few short years and a few million very sparsely distributed sentences to work with.

3 The Tipping Point

The necessity of rules in grammar raises more questions than provides answers. Over the years I have often been asked why, if the "ought" rule is real and applies to seven irregular verbs, we don't see any instance of *hatch-haught* or *wink-wought* along the lines of *catch-caught* and *think-thought*.

Productivity has been extensively studied in the history of linguistics; see Bauer 2001 for an encyclopedic review. But let's start with the intuition shared across all approaches. Productivity concerns the validity of a generalization when novel circumstances arise. We say that the "add -*d*" rule in English past tense is productive because it readily applies to novel verbs with predictable outcomes. By contrast, the "ought" rule is not productive, because it does not apply beyond a finite list of items that must be learned by fiat. In the derivational domain, the nominalizing suffix -*er* productively applies to verbs to generate forms such as *kick-kicker*, *dream-dreamer*, and *blog-blogger*. By contrast, the -*age* suffix seems lexicalized, because it produces derived forms with idiosyncratic meanings (e.g., *marry-marriage*, *carry-carriage*; Chomsky 1970). Indeed, funny things would happen if we were to treat -*age* like -*er*, for we would be looking very hard for the meaning of *sause* in *sausage* or try to make sense of the *mess* in *message*.

The aim of this chapter is to develop a calculus, the Tolerance Principle, that provides a formal and rigorous resolution to the productivity problem. My approach is to tackle the problem from the perspective of a child learner, only making use of plausible psychological mechanisms and cognitive resources available in language acquisition; the OED should not be a prerequisite for acquiring the morphology of English. The principal motivation comes from the complexity of language use. I argue that as the number of exceptions to a rule increases, the real-time processing cost associated with the rule also increases accordingly but in a superlinear trajectory: at some point, it becomes more efficient to lexically list everything and there will be no productive rules.

This will be a long argument. We need to establish all facets of the Tolerance Principle, especially the assumptions about language processing from which the threshold for productivity can be derived. Let's start with the first step: How do children extract generalizations from their linguistic experience?

3.1 Learning by Generalization

All linguistic rules and generalizations can be stated as follows:

(1) R: IF A THEN B

where *A* provides the structural description for *R* which, if met, triggers the application of *B*, the structural change. Quite generally, learning can be framed as a search problem that identifies the structural descriptions of the items that undergo a specific structural change. All explicit learning models in the study of language (e.g., Albright and Hayes 2003; Berwick 1985; Chomsky 1955; Skousen, Lonsdale, and Parkinson 2002) as well as from adjacent fields (artificial intelligence: e.g., Cohen 1995; Daelemans et al. 2009; Mitchell 1982; Yip and Sussman 1997; cognitive psychology: e.g., Feldman 2000; Osherson and Smith 1981) converge on a shared insight: inductive learning must proceed conservatively, drawing minimal generalizations from the data. For concreteness, I illustrate the inductive process using the Yip-Sussman model (Yip and Sussman 1997 and extended by Molnar 2001) on the familiar example of English past tense.

The learner constructs rules as mapping relations between input (stem) and output (past).[1] The input and output are represented as a linear sequence of phonemes specified by their distinctive features in the Yip-Sussman model but I will use English orthography for ease of presentation. The operation of the model is presented in Figure 3.1 with a sequence of input words that becomes incrementally available to the learner.

At the start of learning, there are no rules. Suppose now the first item *walk-walked* comes in, which allows the learner to detect the *-d* suffix.[2] No generalization is possible on a single piece of data, and a trivial rote rule is learned: "IF *walk* THEN *-d*." Suppose the next item is *talk-talked*. As before, the learner detects "IF *talk* THEN *-d*" but now generalization can take place: the two rules constructed so far can be collapsed into one because they involve identical structure change. The learner proceeds to discover the (partial) similarity between *walk* and *talk*: a conservative generalization yields that, for instance, they differ only in the first consonant, which then must not restrict the applicability of the structural change. An intermediate rule is formed: "IF ⋆*alk* THEN *-d*," where ⋆ is a wild card that stands for any segmental material. That is, at the present moment, any stem with the rime *alk* can add *-d* irrespective of the phonological material in the onset, and the previous two separate statements

---

1 Again, the availability of such pairs is by no means an innocuous assumption (section 2.1.1) and readers are directed to the models of Chan (2008) and Lignos et al. (2009) to see how the learner may identify input-output pairs in an unsupervised setting.

2 For clarity of presentation, I also abstract away from the phonological realization of *-d*, which would be /t/ in the case of *walked*. As will be shown later, the Yip-Sussman model is capable of learning the voicing feature agreement between the final segment of the stem and the suffix.

3.1 Learning by Generalization   43

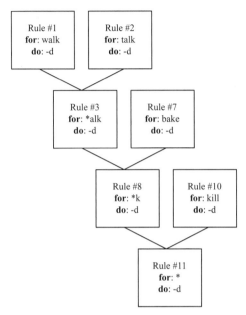

Figure 3.1
The learning of the regular rule ("-*d*"). (Adapted from Molnar 2001.)

can be discarded.³ As more items are presented incrementally, the condition for the application of -*d* becomes broader. Eventually, the learner concludes that -*d* has no restrictions whatever (⋆'s all around as in Rule 11 in Figure 3.1), which corresponds to the notion of a default rule: anything goes. In the present learning model, then, the generality of a rule is directly related to the diversity of words it applies to.

Upon the presentation of English words in the stem and preterite form, the Yip-Sussman model can induce very accurate rules of English morphophonology that are similar to Bloch's (1947) formulation. The rules in (2) are typical output of the model on a few dozens examples of regular verbs: they are presented in terms of distinctive features as implemented in the computational

---

3 Or not. Implicit in the present discussion, and also a logical possibility as well as an empirical claim, is that learners treat regularly suffixed words (e.g., *walked*) on par with the irregulars before they discover the productivity of the "-*d*" rule. Once its productivity is established, the "irregular" variant of *walked* may be eliminated in favor of economy, but it perhaps will also linger on in the lexicon along with its productively composed twin *walk-d*. This may account for the purported whole-word storage effects of highly frequent regulars (Alegre and Gordon 1999; Baayen et al. 2003), although see Lignos and Gorman 2012 for a reinterpretation.

model. These rules clearly correspond to the rules of English regular past tense, where the phonological shape of the suffix is conditioned on the final segment of the stem.

(2) ∅ → d / [+sonorant] __
 ∅ → d / [+voice, −coronal] __
 ∅ → t / [−voice, +strident] __
 ∅ → əd / [+coronal, +anterior, −continuant] __

When trained on the irregular verbs, the Yip-Sussman model induces rules much like those found in traditional descriptions of the English past-tense system.

(3) a. Rime → ɔt/ __ (e.g., *think-thought, catch-caught, buy-bought*)
 b. ɪ → æ / __ ŋ (e.g., *sing-sang, ring-rang*)
 c. d → t / en__ (e.g., *bend-bent, lend-lent, spend-spent*)
 d. [+high, +ATR] → o / __ z (e.g., *choose-chose, freeze-froze, rise-rose*)
 e. i → e / __ [d, t] (e.g., *feed-fed, lead-led, meet-met*)

It remains an open question whether the Yip-Sussman model and similar approaches accurately reflect the cognitive process of language acquisition. To get the model (Figure 3.1) off the ground, children must be able to operate across several levels of linguistic analysis: at a minimum, they need to recognize the semantics and pragmatics of past tense, the mapping between syntactic structures and word formation, and the bundles of phonological features that are the raw material for generalization. And the true quality of a learning model can only be assessed against the reality of child language acquisition. Here we see the deficiencies of the Yip-Sussmann model, because it misses empirical generalizations that can only emerge if the model is enriched with additional structural assumptions and constraints. For instance, irregular verbs such as *bleed-bled* and *feed-fed* in (3e) are subsumed under the vowel shortening process (Halle and Mohanan 1985; Myers 1987), a type of readjustment rule (Halle and Marantz 1994) that is (lexically) triggered under suffixation. That is, *come-came*, *flee-fled*, and *sleep-slept* should be treated in a broader equivalent class even though they take different suffixes (-ø, -d, and -t). This treatment is further supported by the evidence from child language: as reviewed in section 2.2, all vowel-shortening irregulars are learned extremely well regardless of their past-tense frequency, suggesting that they are collectively learned as a group with mutually reinforcing benefits. The Yip-Sussman model, however, has no means of detecting such abstract generalizations; the three classes of vowel shortening irregular verbs are listed separately under distinct rules.

Critically, the rules in (2) and (3) solve only half of the learning problem. For example, the rule "Rime → ɔt/ __ ," as stated, will map the past tense of numerous verbs to /ɔt/. It overreaches because the verbs presented in the input have wide-ranging phonological properties (e.g., *bring, buy, catch, fight, seek, teach, think*), which will lead inductive learners to conclude that the structural description is wide open and unrestricted (wild card all around in the Yip-Sussman formulation), and the past tense of *love* should be *lought*, *reek* would become *rought*, etc. This is clearly catastrophic. And as we have seen in chapter 2, children never overgeneralize the irregular rules. Inductive-learning models are quite good at detecting the *form* of rules but have serious deficiencies when it comes to the *scope* or productivity of rules.

In sum, we still don't fully understand the mechanics by which children derive rules and generalizations from data. Fortunately, children's language development once again guides our study of learning and productivity. For example, the overregularization of the English past tense suggests that the emergence of productivity is not instantaneous but relies on the cumulative effect of experience. "Adam" (Brown 1973), the poster child for past-tense acquisition (Pinker 1995, 1999), showed no errors of overregularization at all prior to 2;11, when *What dat feeled like?* was produced. Of course, Adam would have encountered plenty of regular verbs before that but he evidently resisted elevating the *-d* suffix to the default status until then. Such findings provide valuable clues on how children calibrate productivity; by assessing their vocabulary, we also can evaluate the numerical balance between rules and exceptions. The Tolerance Principle, as we will see, provides a precise answer for how many regular verbs will be needed to trump the irregulars.

3.2 The Cost of Exceptions

Imagine a point in language development where a child's entire vocabulary consists of two verbs: *know-knew* and *grow-grew*. He or she might be tempted to posit the "ow-ew" rule (o → u) following some inductive-learning scheme such as the Yip-Sussman model. At this very moment, the child has every reason to believe the "ow-ew" rule to be fully productive, for it is 100% compatible with his or her linguistic experience. But as the child's vocabulary expands, more exceptions will join in: *sow-sowed, show-showed, tow-towed, snow-snowed*, and so on, even though the rule may also pick up supporting members (*blow-blew, throw-threw*). In any case, since there are only a finite number of irregular verbs that follow "o → u," sooner or later the rule will be thoroughly outnumbered by the exceptions and meet its demise. (More precisely, the rule will become unproductive, applying to a fixed list of items but not beyond.) In contrast, the some 150 irregular verbs in spoken English apparently aren't quite enough to thwart the productivity of *-d*, which is supported by thousands of regular verbs.

The intuition behind the "worthiness" of rules is exactly what Mark Aronoff (1976, 36) expresses in his classic treatment of productivity. For a word-formation rule (WFR),

> We count up the number of words which we feel could occur as the output of a given WFR (which we can do by counting the number of possible bases for the rule), count up the number of actually occurring words formed by that rule, take the ratio of the two and compare this with the same ratio for another WFR.

The score for the "ow-ew" rule is two out of two when our hypothetical child only knows two verbs. And it's quite clear that this ratio will go down as learning proceeds. What's left open is the precise batting average that warrants productivity: How many exceptions would topple "ow-ew"? How many irregular verbs can the "*-d*" rule sustain? In his pioneering study, Aronoff came to ponder the very same question: numerical factors appear to be a primary motivation for the establishment of morphological relations. For instance, when considering the derivation of English adjectives of the form *Xistic* as in *imperialistic*, Aronoff (1976, 118) investigates whether the correspondence can be established on the basis of nouns of the form *Xist* (*imperialist*). If this derivation is productive — that is, a $Xist_N$ is the source of $Xistic_A$ — then a $x_i ist_N$ should exist for almost every $x_i istic_A$. Consulting a dictionary (Walker 1936), Aronoff finds 145 words of the form $Xistic_A$, but 28 of them, such as *characteristic* and *logistic*, do not have a corresponding $Xist_N$ form: "too many exceptions to our proposed derivation for it to be above suspicion" (118). Similarly, Aronoff notes that for the 145 $Xistic_A$ items, most have a corresponding $Xism_N$ form (*imperialistic-imperialism*), with 26 exceptions such as *stylistic*

3.2 The Cost of Exceptions   47

and *linguistic*. Alternative formulations were pursued because the numbers of exceptions are regarded as too high.[4]

The quantitative study of productivity has a checkered history in linguistics and psycholinguistics: multiple paths of investigation are possible but they need to be placed under the light of language acquisition. For instance, a long line of research from Baayen and colleagues aims to develop numerical measures of productivity that would "accord nicely with [linguists'] intuitive estimates of productivity" (Baayen and Lieber 1991, 801). Albright (2002) uses lexical statistics to correlate with speakers' rating responses to conceivable inflectional forms of novel words. And recent work has seen more sophisticated statistical analyses that incorporate psycholinguistic and sociolinguistic findings (Plag and Baayen 2009; Plag, Dalton-Puffer, and Baayen 1999). But it is important to recognize these approaches as statistical summaries of data, rather than learning models that make productivity decisions. Obviously child learners have no access to their caretakers' grammaticality ratings, or mixed-effects regression over corpora, or lexical decision time measurements during the course of language acquisition. What they do have is a very sparse collection of data that consists of mostly very simple words. Furthermore, even if a statistical analysis of productivity in many empirical cases turned out to be correct—say, a productive rule can tolerate 20% of exceptions—it would still provide no insight on the psychological reality of productivity, or why the critical value should be 20%, rather than 18% or 26%. In fact, a moment of reflection tells us that a purely data-drive approach of productivity cannot in principle succeed. This is because for productive processes, the number of exceptions may be far below the critical value, and for productive processes, far above. No amount of statistical regression can identify the precise criterion for productivity.

The approach I develop here is a throwback to the notion of *evaluation metrics* proposed in the founding documents of generative grammar (Chomsky 1955, see also Chomsky 1965 as well as Chomsky and Halle 1968). The evaluation metric is conceived as a theoretical device that guide linguists to choose among competing analyses and enables children to select the correct grammar on the basis of the primary linguistic data. As Chomsky (1965) stresses, the choice of an evaluation metric "is not given a priori ... Rather, a proposal

---

4 It seems that Aronoff was too conservative. According to the Tolerance Principle, a value of $N = 145$ can tolerate $\theta_{145} = 29$ exceptions: the derivation from $Xist_N$ to $Xistic_A$ and from $Xistic_A$ to $Xism_N$ *can* be productively maintained.

concerning such a measure is an empirical hypothesis about the nature of language" (p. 37), and the notion of simplicity is not "a general notion somehow understood in advance outside of linguistic theory" but "an empirical matter with empirical consequences" (p. 38; see also Sober 1975). As a proposal for child language acquisition, then, the Tolerance Principle must be rigorously justified as an empirical hypothesis.

I propose that the calibration of productivity minimizes the computation of rules and exceptions. In general, there are two measures — space and time — that formally figure into complexity considerations. Both metrics are in principle valid for the study of language and the choice must be based on their empirical merits. There has been growing interest in the distributional learning of language, often couched in the *Minimum Description Length* (MDL) framework (Cover and Thomas 2012; Rissanen 1978) or other formally equivalent approaches (Chater and Vitányi 2007; Tenenbaum and Griffiths 2001); see Goldsmith 2001 for a general introduction with specific focus on morphology and Hayes and Wilson 2008 for an application to phonology. This framework views the grammar as a data compression device (Chomsky 1951). It strives to minimize the structural description of data, to eliminate redundancies, and to obtain the simplest and most elegant statement of the grammar — not unlike how linguists construct theoretical analyses. For instance, the regular "-d" rule obviously contributes to the economy of storage such that thousands of regular verbs needn't be individually listed for past tense. One can conceivably devise a scheme such that the postulation of a rule is justified when it achieves more space saving than lexical listing. At the present time, however, an MDL approach seems a recipe for ad hocism. Because we understand preciously little about the constraints on linguistic memory and computation, we have no principled basis to evaluate the cost of storing a word, or the cost of storing a rule, or the cost of storing the mapping between words and the rules that apply to them. Without having an independent measure of these quantities, we do not have a well-motivated currency to assess how space can be minimized under different organizations of language. To make matters worse, we have few concrete clues on child learners' computational power, without which we cannot be certain how much compression can be squeezed out of the data. If pursued to the limit, the "M" in MDL may yield highly abstract descriptions of language such that the storage for words is minimized but the derivational complexity of the inflected forms is maximized, not unlike the approach to phonology developed in SPE (Chomsky and Halle 1968). While I do not wish to claim that the pursuit of descriptive economy is incorrect, we currently know virtually nothing about its computational requirement or psychological grounding. As it stands, the MDL approach does not move us closer to an empirical understanding of language acquisition.

If it's not space, then it must be time. The Tolerance Principle provides an evaluation metric that quantifies real time language processing. In particular, I suggest that learners always chooses the more efficient, i.e., faster, organization of word formation. A productive rule is postulated if it speeds things up; otherwise the learner favors lexical listing.

The argument goes as follows (and it's a long one). In section 3.3, I propose, and justify, the Elsewhere Condition as a cognitive model of processing rules and exceptions, drawing on the psycholinguistic literature on lexical and morphological processing. In section 3.4, I provide the derivation of the Tolerance Principle, which is a mathematical consequence of the Elsewhere Condition. In section 3.5, I make some methodological remarks on how the Tolerance Principle, especially its recursive application, may guide children toward the grammar.

3.3 Elsewhere in Language Processing

A traditional approach to rules and exceptions is the Pāṇinian Elsewhere Condition (Anderson 1969; Brown and Hippisley 2012; Halle 1997; Halle and Marantz 1993; Kiparsky 1973; Stump 2001), which is employed in a wide range of linguistic theories. Specifically, exceptions are handled by more specific rules or processes, which bleed the application of the general rule. At the algorithmic level, the Elsewhere Condition can be implemented as a serial search procedure, where the exceptions are treated as conditional statements prior to the application of the general rule:

(4)   IF $w = w_1$ THEN ...
      IF $w = w_2$ THEN ...
      ...
      IF $w = w_e$ THEN ...
      Apply R

In (4), a lexical item w is first evaluated against the listed exceptions.[5] If a match is found (e.g., $w = w_i$), the associated exceptional clause is triggered. If the list has been exhausted without finding a match, then the rule R applies. The key claim of the Elsewhere Condition model is that the computation of rules and exceptions is serial, following an earlier tradition in psycholinguistics (Forster 1976; Sternberg 1969) as opposed to the currently more popular associative accounts of lexical processing (e.g., McClelland and Rumelhart 1981; Plaut 1997). As we will see, the search for exceptions prior to the application of the rule contributes to the rising cost of processing as the number of exceptions increases.

The serial nature of the Elsewhere Condition can be interpreted as a component of the competence theory, similar to the invocation of computational efficiency in the Minimalist Program (e.g., the Minimal Link Condition, probe-goal search; Chomsky 1995, 2001). Here I would like to go a step further. I propose that the Elsewhere Condition as embodied in (4) is simultaneously a model of linguistic performance, again reviving an earlier tradition in linguistics and psychology (Fodor, Bever, and Garrett 1974). As Miller and Chomsky (1963, 481) put it, "The psychological plausibility of a transformational model of the language user would be strengthened, of course, if it could be shown that our performance on tasks requiring an appreciation of the structure of transformed sentences is some function of the nature, number, and complexity of the grammatical transformation involved." This somewhat unconventional argument is constructed from several independent lines of psycholinguistic investigations that I now take up in turn.

3.3.1 Listing Exceptions

Consider a list of exceptions such as the irregular verbs of English. The Elsewhere Condition claims that they form a list that is traversed in a sequential fashion, and more specifically, in decreasing order with respect to their frequency. Thus, a more frequent item will be accessed faster than a less frequent item because it is placed in a higher position on the list and will be accessed sooner.

---

[5] For present purposes, I set aside whether the derived form of an exceptional item (e.g., *catch*) is retrieved from memory (Pinker 1999: "fetch *caught*") or processed by a lexical/irregular rule (section 2.2; "apply Rime → ɔt/"). The main point is that exceptions are handled by more specific processes, prior to the application of the general rule.

The serial search of exceptions straightforwardly accounts for the well-established frequency effects in the processing of irregularly formed words: higher-frequency irregulars tend to be processed faster than lower-frequency irregulars in both recognition and production tasks (for reviews, see Clahsen 1999; Marslen-Wilson and Tyler 1997; Pinker and Ullman 2002).[6] Additionally, frequency effects receive a very specific interpretation under the Elsewhere Condition: the operation of serial search implies that it is the *rank*, or relative frequency, of words that determines the speed of irregular processing. Rank is of course dependent on absolute frequency but it is possible to contrast them in studies of lexical processing (Murray and Forster 2004). For such a comparison, I turn to a large-scale lexical decision study, the English Lexicon Project (ELP; Balota et al. 2007), which collected almost three million reaction-time measurements for over forty thousand words. I extracted all the reaction-time data for irregular past tense, excluding the no-change verbs such as *hit* since they are indistinguishable from the infinitive form, as well as morphologically derived irregulars (e.g., *rewear-rewore*). This results in ninety-seven verbs. Their lexical decision time is significantly correlated with rank, which provides a slightly better fit of the logarithm of lexical frequency (r: 0.67 vs. 0.64, adjusted R^2: 0.44 vs. 0.40). Therefore, the serial search process in the Elsewhere Condition provides as good an account of the lexical decision time of irregular processing as the more conventional frequency-based explanations.

Rest assured that maintaining and using a frequency-ranked list is well within the means of human language processing. Such a list needn't be constructed by keeping track of all the frequencies and then rearranging them in descending order. The serial search approach to lexical process is partly motivated by online algorithms designed for frequency-sensitive computation. A simple implementation is the "move to front" algorithm: whenever a word is used, it is moved to the beginning of the list. Another elegant model is the "move up" algorithm (Rivest 1976), which swaps the item just used with the one ranked just above it. Both algorithms are computationally trivial but are

---

[6] Frequency effects in irregular processing have traditionally been viewed as evidence for the associative retrieval of holistically formed words. But the facts are also consistent with a morphological processing model under which irregulars are compositionally formed as well (see, e.g., Allen and Badecker 2002, Stockall and Marantz 2006, Taft 2004 and especially Lignos and Gorman 2012).

known to be near optimal (Sleator and Tarjan 1985a).[7] Readers only need to look at their smartphones to be convinced. Double tapping the home button on an iPhone will reveal the list of active apps sorted by recency — which closely matches their relative frequency of usage.

Several well-known findings in lexical processing follow immediately from the serial search model; associative models appear problematic or more complex (Forster 1992). For instance, on use or exposure, a word will be moved to a higher position on the list. This will result in faster retrieval in subsequent tasks, providing a straightforward explanation for recency effects in lexical processing (e.g., Scarborough, Cortese, and Scarborough 1977) Furthermore, the serial search model, under which it is the relative rank of words rather than the frequency of occurrence that determines the speed of lexical access, immediately accounts for the general absence of age effects in lexical processing. For example, Cerella and Fozard (1984) find no difference between twenty-three-year-old and seventy-three-year-old subjects' reaction time for word-naming and semantic priming tasks; see also Balota and Duchek 1988. Presumably, the ranks of words are comparable across individuals, while the fifty additional years of experience with words — thus higher word frequencies for older people — evidently do not speed up lexical access, which is unexpected under associative accounts that take lexical frequency, rather than relative rank, as the determinant factor in the speed of processing.

3.3.2 Exceptions before Rules

The second, and more critical, claim of the Elsewhere Condition model concerns the processing of the rule-following items (the "regulars"). Specifically, I assert that in order to inflect a regular verb such as *walk*, the language processor must first make sure the stem is not one of the listed exceptions prior to the application of the "-*d*" rule. In other words, the regulars must wait for the irregulars.

This seems absurd. One can certainly design a more efficient computational system that bypasses the Elsewhere Condition. For instance, the language user may create a marker for the regular verbs, which immediately triggers -*d* suffixation without concerning the irregulars at all. But it would be extremely

---

7 Similar algorithms with more sophisticated data structures may be used for finer-grained models of the mental lexicon. For instance, it may be interesting and fruitful to explore the Cohort effects in lexical access (Marslen-Wilson 1987) with lexicographic tree search algorithms (Sleator and Tarjan 1985b).

3.3 Elsewhere in Language Processing

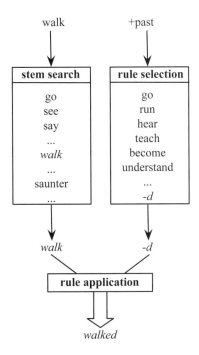

Figure 3.2
The past-tense inflection of *walked* with the three independent components of morphological computation

interesting, and indeed surprising, if the online computation of words has no choice but to follow the puzzlingly cumbersome principle of the Elsewhere Condition, as I proceed to demonstrate.

In general, the inflection of morphologically derived words involves at least three processes (Caramazza 1997; Fruchter and Marantz 2015; Levelt, Roelofs, and Meyer 1999; Taft 2004; Taft and Forster 1975):

(5) a. Stem search: lexical lookup of the stem (e.g., *think*, *walk*).

   b. Rule selection: search for the appropriate morphological rule(s) for the stem (e.g., the irregular "ought" rule for *think* and the regular *-d* rule for *walk*).

   c. Rule application: use the selected rule to generate the inflected form (e.g., *think* → *thought*, *walk* → *walked*).

Figure 3.2 illustrates the three components in the inflection of *walk-walked*.

The stem search and rule selection processes in (5) are both serial and sensitive to lexical rank effects, and rule selection additionally follows the Elsewhere Condition. That is, the activation of the "-*d*" rule only takes place after the algorithm has rejected the irregular verbs (w_1, w_2, ...), which fail to match the target stem *walk*. The rule application component then generates the inflected form. Had the stem been an irregular verb, the rule selection search will immediately terminate when a match (w) is found and the correspondingly irregular rule will be retrieved. In principle, the three components are executed independently, and in general, they must be. To process morphologically complex languages such as Turkish and Finnish, the holistic storage of derived forms, which number in the billions (Hankamer 1992; Niemi, Laine, and Tuominen 1994), seems quite implausible, especially under the data sparsity considerations reviewed in chapter 2.

The Elsewhere Condition corresponds to the rule selection component (5b) in the general modeling of morphological processing; that will be our focus momentarily but first the other two components require justification as well.

3.3.2.1 Finding Stems and Using Rules

The stem search process follows frequency effects widely known since the 1970s (Taft 1979). Consider two words *pleasing* and *indexing* that are approximately matched in word frequency (cf. Taft 2004). The lexical decision time data from the English Lexical Project (Balota et al. 2007) shows that *pleasing* is recognized a whopping 287.7 ms faster than *indexing*, an eternity in lexical decision. The advantage for *pleasing* is attributed to the stem search process. The stem *please* is far more frequent than the stem *index* (595,358 vs. 37,725 in the frequency counts provided in the ELP study), where the frequency is tallied over all lexemes that contain the stem (i.e., from *please*, *pleased*, *displease*, etc.). Stem frequency effects formed the original motivation for serial search models of lexical and morphological processing (Taft and Forster 1975, 1976).

The rule application component is in action when morphologically complex words are assembled online: the generation of *warned* by attaching *-d* to *warn* is not instantaneous and takes time, in what Taft (2004) refers to as the recombination stage. At this juncture, we may wonder if rules are differentiated in their speed of online processing. For instance, the *-d* past-tense suffix is considerably more frequent than the third-person singular present tense suffix *-s*: collectively, the *-d* suffix roughly doubles the *-s* suffix in total frequency (Balota et al. 2007) when counting only items that are exclusively verbal forms.[8] Now *warned* and *warns* share the same stem and happen to have very similar surface frequencies: a strict interpretation of the morphological processing model in (5) predicts that *warned* would be processed faster than *warns*, thanks to its more frequent suffix/rule.

A suffix/rule frequency effect is just the logical extension of the stem frequency effect. Most theories of morphological processing (Caramazza, Laundanna, and Romani 1988; New et al. 2004; Stockall and Marantz 2006; Taft 2004; Taft and Forster 1975) share the assumption that stems and affixes have independent representations and are combined online in real time, at least for the units that participate in productive word formation. While stem frequency effects have been extensively documented, I am only able to find one study that specifically targets the effect of suffix frequency (Colé, Beauvillain, and Segui 1989). Again, the English Lexicon Project provides a fertile testing ground. I compare the processing time of verbs with the *-d* and *-s* suffixes. Only unambiguously verbal forms are selected. Furthermore, I only compare word pairs such as *warned* and *warns* that (a) share the same stem, thus stem frequency, and (b) have very similar surface frequencies, thereby controlling for the transitional probabilities between the stem and the suffix that seem to affect the time course of morphological segmentation and composition (e.g., Baayen et al. 2003; Solomyak and Marantz 2010). These strict criteria yield relatively few (thirty) pairs of past-tense and third-person singular verbs, and their lexical decision time are subjected to a pairwise comparison. The surface frequencies of these pairs are closely matched (mean frequency 4,885 and 4,453 in Balota et al. 2007; difference not significant, $p = 0.24$). A paired Mann-Whitney test shows a clear advantage for the *-d* suffixed verbs over the *-s* suffixed verbs: the average speedup is 21 ms ($p = 0.018$). This suggests that the rule application process, like stem search, is sensitive to the frequency of the morphological constituent.

In fact, the speedup due to suffix frequency may be ever stronger. Lexical decision is, by hypothesis, a composite of all three factors listed in (5). Although pairs such as *warned* and *warns* are matched for both stem and surface frequencies, they not only differ in suffix frequency but with respect to the rule selection process. According to the Elsewhere Condition (4), both inflections need to scan through the list of exceptions before reaching their respective

---

8 That is, the *-d* in *tired* and *-s* in *barks* are not included due to the category ambiguity of these words.

suffix. But here the regular -*d* forms are at a disadvantage because of the irregular verbs, which need to be searched and rejected. The -*s* suffix, by contrast, is exceptionless and the algorithm can select the rule without delay. Despite this, the -*s* suffixed verbs are still slower than the -*d* suffixed verbs, which reinforces the suffix frequency effect in the rule application process.

3.3.2.2 Selecting Rules

Having established the independent effects of stem search and rule application, I examine the role of rule selection in the Elsewhere Condition model, the centerpiece in our proposal of morphological computation.

For concreteness consider two stems w_e and w_r. Both stems follow the structural description of R, but R only applies to w_r by adding a productive suffix s_r, whereas w_e is an exception that takes a different/exceptional suffix s_e. Suppose further that w_r and w_e have these additional properties:

(6) a. They are matched in stem frequency: $f(w_e) = f(w_r)$.
  b. Their derived forms ($w \cdot s$) are matched in surface frequency: $f(w_e \cdot s_e) = f(w_r \cdot s_r)$, and thus the stem-suffix transitional probabilities are also matched: $P(w_e \to s_e) = P(w_r \to s_r)$.
  c. The suffixes are matched in rule frequencies, which are the sum of the frequencies of all derived words taking the suffixes: $f(s_e) = f(s_f)$, where $f(s) = \sum_w f(w \cdot s)$.

That is, the exceptional and regular stems are completely matched for all the relevant factors that affect processing, and the only remaining difference concerns the selection of the rule/suffix. According to Elsewhere Condition model, the s_e suffix will be selected prior to the s_r suffix due to the exception-before-rule nature of the serial search. I thus predict that the exceptional form ($w_e \cdot s_e$) will be processed faster than the productive form ($w_r \cdot s_r$).

Here *ceteris paribus* is the key: I do not claim that regulars are always slower than irregulars but only when stringent conditions are met to neutralize other factors. But all things are rarely equal. The English morphological system, unfortunately, does not provide a suitable testing ground. While it is possible to find regular and irregular verbs matched in both stem and surface frequencies, their suffix/rule frequencies, which I have just shown to affect the speed of morphological processing, are very different, making it impossible to isolate and test the effect of rule selection. Specifically, while some of the most frequent verbs are irregular, none of the irregular rules come anywhere near the suffix frequency of -*d*, which accounts for 57.2% of all past tense tokens (the CELEX corpus; Baayen, Piepenbrock, and Gulikers 1996).

Fortunately, the morphology of German provides suitable word pairs to test the Elsewhere Condition as an online processing model. The most direct evidence comes from the past participle system, which makes use of three morphological processes: prefixation, suffixation, and stem allomorphy.

(7) a. kaufen ge-kauft
'to buy' 'bought'

b. saufen ge-soffen
'to booze' 'boozed'

c. brennen ge-brannt
'to burn' 'burned'

d. laufen ge-laufen
'to run' 'run'

e. verlaufen verlaufen
'to go astray' 'gone astray'

The verb stem is prefixed with *ge*- if it bears initial stress (Kiparsky 1966), as can be seen in the contrast between (7d) and (7e). The suffix is always one of two segmentable endings, -*t* and -*n*. There is considerable evidence that the -*t* suffix is the productive default. For instance, when entering the German lexicon, novel verbs such as *simsen* ('to send an SMS') automatically take the -*t* suffix (see Marcus et al. 1995 and Clahsen 1999 for reviews). The -*n* suffix, however, attaches to a synchronically unpredictable list of stems. For instance, *kaufen, saufen* and *laufen* in (7) are phonologically similar but take different suffixes, and the irregular -*n* suffix may also trigger stem change as in *saufen-gesoffen*.

The -*t* and -*n* suffixes have very similar collective frequencies: according to the German CELEX corpus, 46.89% of the participle tokens are regular (Westermann 2000; see Clahsen 1997 for comparable ratios based on child-directed German). That is, the regular -*t* and the irregular -*n* are very closely matched in suffix frequency. If we find irregular and regular verbs with matched stem and surface frequencies, the irregular participles are predicted to be faster thanks to the rule selection process in the Elsewhere Condition that gives precedence to the irregulars. This prediction is confirmed in a spoken production study by Fleischhauer and Clahsen 2012; see also Clahsen, Eisenbeiss, and Sonnenstuhl 1997, Clahsen, Hadler, and Weyerts 2004. Child and adult participants listened to a sentence accompanied by pictures and were instructed to correct a cartoon figure's ungrammatical use of the past particle. Production latency is measured from the end of the cartoon figure's speech to the onset of the participant's speech. I focus on the results involving verbs (irregular and regular) in their high-frequency group, because these are likely to be acquired earlier in language acquisition and thus form the vocabulary over which morphological productivity is calibrated. The irregular *-n* suffixed participles are produced considerably faster than the regulars, with a difference of 37 ms for adults ($p < 0.01$) and 105 ms for children ($p < 0.0001$).[9]

Another strand of evidence can be found in the German noun plural system (Marcus et al. 1995), which employs five plural suffixes (*-s, -(e)n, -e, -er,* and *-∅*). Under many analyses of German morphology (e.g., Hahn and Nakisa 2000; Wiese 1996; Wunderlich 1999), *-s* as in *Auto-Autos* 'car(s)' is productive, whereas the *-er* suffix as in *Ei-Eier* 'eye(s)' or *Wald-Wälder* 'forest(s)' is lexically idiosyncratic. The other three suffixes are controversial: the dual-route model holds them to be unproductive and lexical (Clahsen 1999; Marcus et al. 1995) but most other scholars consider them compositionally formed albeit with exceptions; I return to these intricacies in German plurals in chapter 4. More pertinent for present purposes, note that *-s* and *-er* suffixes have comparable frequencies: both are rare and take up 2% and 3% of the noun plural tokens in the German CELEX database. Thus, *-s* and *-er* suffixed nouns also make suitable materials for testing the predictions of the Elsewhere Condition as a processing model. Sonnenstuhl and Huth (2002) matched the stem frequencies of *-er* and *-s* nouns, and created two groups of high and low surface frequency. In both groups, a significant advantage for the *-er* suffixed over the *-s* suffixed plurals is found in lexical decision: 35 ms for the low surface frequency group and 89 ms for the high surface frequency group. This again supports the exceptions-before-rule scheme in the Elsewhere Condition.

Looking beyond the domain of morphology, it is interesting to consider the general problem of how exceptions are processed in relation to the productive component of the grammar. Structurally, the Elsewhere Condition appears

---

9 I thank Elizabet Fleischhauer for providing the statistical results here. Their published paper (Fleischhauer and Clahsen 2012) only reports mean production latencies because it addresses a different theoretical question.

implicated in the organization of the lexicosemantic component of the grammar and its acquisition. For instance, classifer systems appear to be organized around semantic properties of nouns, and more specific matches are generally preferred over general ones. Examples from Chinese (Wu, Cheng, and Pan 2014) and Japanese (Downing 1996) are provided below.

(8) a. yi TOU zhu/niu
       one TOU pig/cow (domesticated animal)

       yi   ZHI yezhu/yeniu/niao/hu
       one ZHI boar/bison/bird/tiger

    b. -wa: birds
       -too: large animals
       -hiki: (small) animals

Interestingly, the acquisition of classifiers in these languages (Hu 1993; Yamamoto and Keil 2000) shows comparable patterns to the acquisition of morphology as reviewed in chapter 2. When a more specific classifier is needed, children occasionally overuse a more general one. But when a more general classifier is needed, a more specific classifier, or one from a wrong semantic class (e.g., a bird classifier for an inanimate object), is almost never used, showing the same kind of asymmetry observed in English past tense: Overregularization, yes, but overirregularization, no.

Thus, there is evidence for the Elsewhere Condition as a fundamental principle of linguistic organization where specificity and generality come into conflict. This naturally leads us to verify its effects on the time course of rule and exception processing in other linguistic domains. There is considerable research that compares the processing of figurative and literal meanings of an idiom such as *kick the bucket* (i.e., to die vs. the action of kicking), as well as the contrast between idiomatic and fully compositional phrases (e.g., *kick the bucket* vs. *lift the bucket*). In both cases, a real-time advantage for idioms has been observed (Bobrow and Bell 1973; Cacciari and Tabossi 1988; Swinney and Cutler 1979). For instance, Swinney and Cutler asked subjects to read strings of words, some idiomatic and some literal, and decide as quickly as possible if they are meaningful phrases in English. Their key manipulation is to substitute the critical word in both phrases (e.g., from *kick* to *lift the bucket*). A speedup of about 100 ms for the idiomatic expression is consistently observed across all idiom types (e.g., Fraser 1974).[10] These results, which have been corroborated in later studies of both comprehension and production (Gibbs 1980; Gibbs and Gonzales 1985; Tabossi, Fanari, and Wolf 2009; Van Lancker, Canter, and Terbeek 1981), are consistent with the use of the Elsewhere Condition in a general model of language processing. When people assign a structural description (e.g., a syntactic parse or a morphological segmentation) to an expression, they look up in the mental lexicon to see if an exceptional form or meaning can be located; if not, then the compositional interpretation of the expression takes place.

It must be said that direct evidence for the Elsewhere Condition as a performance model is still tentative. The numerous interacting factors in language processing make it very difficult to isolate the mechanism by which speakers tread the line between rules and exceptions in real-time language use. Moreover, we are probably approaching the limit of lexical decision and other conventional behavioral methods for the psycholinguistic study of word formation, and new approaches may be needed (Fruchter and Marantz 2015). Nevertheless, the evidence reviewed here seems to provide a solid footing for the Elsewhere Condition as a model of language processing, which leads to the development of the Tolerance Principle as a calculus for productivity.

3.4 The Tolerance Principle

I conjecture that rules and exceptions are organized to optimize/minimize the time complexity of language use.

Suppose that a rule R may in principle apply to a set of N lexical items. For instance, R may be of the form "IF X is a verb THEN add -d for past tense." Of the N items, a subset of e items are exceptions that do not follow R: the irregular verbs in this example. If R is productive, then the exceptions are arranged on a frequency-ranked list and are accessed before the items that follow R; see (9a). If R is unproductive, however, then all N items would be organized on a lexicalized list, again sorted by their frequencies; see (9b).

(9) a. A productive rule with exceptions:

- w_1

---

10 Interestingly, Swinney and Cutler show that the phrasal transitional probability does not appear to affect the speed of processing: the relatively high frequency of the idiomatic expressions is therefore not the cause of the shorter reaction time in processing.

3.4 The Tolerance Principle

- w_2
- ...
- w_e
- Apply R ($N - e$ items)

b. A nonproductive that lists everything:

- w_1
- w_2
- ...
- w_{N-1}
- w_N

Let $T(N,e)$ be the expected time of rule access if R is productive with e exceptions (9a). Specifically, all items that follow R (the regulars) will have to wait until all the e exceptions are evaluated and rejected. Thus, every regular item consumes e units of time, the number of exceptions. By contrast, for an exceptional item, the rule search time is determined by its rank/position on the list. The overall expected time complexity $T(N,e)$ is then the weighted average of time units over the probabilities of these two sets of items.

Consider the case where R is not productive: all N items must be listed as exceptions, again ranked by their frequencies (9b). Let the expected time of access in this case be $T(N,N)$, where all N items are treated as if they are exceptions. I conjecture that language learners opt for a more efficient model of processing:

(10) *Tolerance Principle*:
R is productive if $T(N,e) < T(N,N)$; otherwise R is unproductive.

The full listing model is optimal with respect to frequency: high-frequency items will be placed higher on the list and thus processed faster. Under a productive rule, all exceptions are listed first, which has the potential effect of demoting a rule-following item to a position lower than its frequency warrants. A sufficiently large number of exceptions may slowdown the overall time complexity to the point where resorting to full listing is more efficient; the Tolerance Principle provides a precise solution for what the tipping point may be.

I now derive the Tolerance Principle. Throughout my analysis, I assume that any sample of words follows Zipf's law (Zipf 1949); this simplifies the calculation and, as we will see at the end of this chapter, has some very interesting consequences when we put the Tolerance Principle into use. Specifically, in a sample of N distinct word types $\{w_1, w_2, ...w_N\}$, the rank (r_i) and frequency (f_i) of the word w_i are inversely proportional — that is, $r_i f_i = C$ for some constant C. The probability of occurrence (p_i) for w_i can be expressed as follows:

$$p_i = f_i / \sum_{k=1}^{N} f_k$$

$$= \left(\frac{C}{r_i}\right) / \sum_{k=1}^{N} \frac{C}{r_k}$$

$$= \frac{1}{iH_N} \text{ where } H_N = \sum_{k=1}^{N} \frac{1}{k}$$

Since the rth-ranked word in a list of N items will be reached with r steps in the serial search (as in 9b), we have

$$T(N,N) = \sum_{r=1}^{N} r \frac{1}{rH_N} = \frac{N}{H_N}$$

Consider now $T(N,e)$. Recall that e is the number of exceptions ranked by frequency: under the Zipfian assumption, the expected time for accessing the exceptions is $T(e,e)$ or e/H_e. For the other $(N-e)$ items, the access time is the constant e, the number of exceptions. Thus, the overall average for the rule-plus-exception model (9a) is

$$T(N,e) = \frac{e}{N} T(e,e) + (1 - \frac{e}{N})e$$

$$= \frac{e}{N} \frac{e}{H_e} + (1 - \frac{e}{N})e$$

Before we derive an analytical solution to $T(N,N) = T(N,e)$ for the variable e, which will give the critical threshold for productivity, consider the relationship between the two quantities in Figure 3.3 based on a numerical simulation.

The dotted line represents the expected search time for a list of $N = 100$ items, or $T(100, 100)$, which is obviously a constant. The solid line represents the expected search time for having a productive rule with an increasing number of exceptions (e), from 1 all the way up to $N = 100$. Figure 3.3 shows that when there are few exceptions (i.e., e is small), it is more economical to scan through them before invoking the productive rule. But as e increases, roughly at the value of $e = 22$, it becomes more economical to have a completely lexicalized list.

3.4 The Tolerance Principle

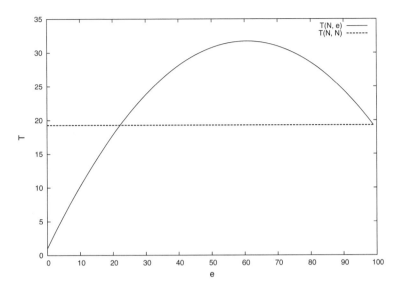

Figure 3.3
$T(N,N)$ vs. $T(N,e)$, where $N = 100, 1 \leq e \leq 100$

With the help of Sam Gutmann, we derive a closed-form solution to $T(N,N) = T(N,e)$. We first approximate $H_N = \sum_{i=1}^{N} \frac{1}{i}$, the Nth harmonic number, with the natural log ($\ln N$). We would like to find $x = e/N$ such that

$$x \frac{e}{\ln e} + (1-x)e = \frac{N}{\ln N}$$

Dividing both sides by N and making use of a fact about logarithm:

$$x^2 \frac{1}{\ln N + \ln x} + (1-x)x = \frac{1}{\ln N}$$

Let

$$f(x) = x^2 \frac{1}{\ln N + \ln x} + (1-x)x - \frac{1}{\ln N}$$

Observe

$$f(\frac{1}{\ln N}) = \frac{(1/\ln N)^2}{\ln N - \ln \ln N} + (1 - \frac{1}{\ln N})\frac{1}{\ln N} - \frac{1}{\ln N}$$
$$= -(\frac{1}{\ln N})^2 + (\frac{1}{\ln N})^3 \frac{\ln N}{\ln N - \ln \ln N}$$
$$\approx -\left(\frac{1}{\ln N}\right)^2$$
$$\approx 0 \quad \text{for large values of } N$$

We thus derive

(11) *Tolerance Principle*

Let R be a rule applicable to N items, of which e are exceptions. R is productive if and only iff

$$e \leq \theta_N \text{ where } \theta_N := \frac{N}{\ln N}$$

Figure 3.4 presents the closed-form solution (θ_N) given by (11) in comparison to simulation results: $N/\ln N$ is a very good approximation of the critical threshold for productivity.

At the present time, it is not clear how the Tolerance Principle is executed as a cognitive mechanism of learning: surely children doesn't use calculators. Conceivably, the calibration of productivity is implemented via the real-time competition of rules and lexical listing, or $T(N, e)$ vs. $T(N, N)$. Learners have two ways of organizing words and exceptions, and the faster route wins and the slower one atrophies. This notion of competition, which determines the outcome of learning, is quite different from proposals in morphological processing (Anshen and Aronoff 1988) that are also competition based. For instance, Baayen, Dijkstra, and Schreuder (1997) suggest that morphological processing always has two routines: holistic storage and compositional formation. The two processes operate in parallel; a word is recognized depending on which of the two processes finishes first. The dual representation was motivated to account for the apparent storage effects of some regularly inflected words (see also Alegre and Gordon 1999) but these results have been called into question (Lignos and Gorman 2012; see also note 3). Regardless, the Tolerance Principle uses speed to select the lexical organization optimized for all words, whereas "race" models use speed to access the favored representation for each individual word.

A final technical point before we consider the implications of the Tolerance Principle for language learning. It has often been observed that exceptions tend

3.4 The Tolerance Principle

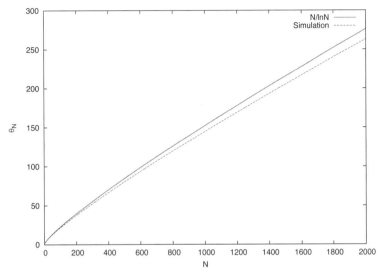

Figure 3.4
The analytical solution for the critical threshold closely matches numerical results.

to clustered the high-frequency region of words. For example, as we shall see in chapter 4, fifty-four of the top one hundred most frequent verbs in child-directed English are irregular. How do such top-heavy distributions of exceptions affect the calculation of the Tolerance Principle, which assumes that the exceptions are scattered randomly among the N lexical items? The answer is, not much. First, the concentration of English irregular verbs at the very top appears to be an outlier in a wide range of empirical cases. For instance, only six out of the top one hundred most frequent nouns have irregular plurals (e.g., *people*, *men*, *children*). Second, it is easy to show that even if all e exceptions are in the top half of the N items, the threshold $\theta_N = N/\ln N$ will hard move. The calculation of $T(N, e)$ requires the probability of encountering an exception, which is approximately $2e/N$ if all e's are in the top half. For each of these e exception, the maximum number of steps the ranked-based search is e. The probability of drawing a rule-following item is thus $(1 - 2e/N)$ with the number of search steps being e as well. Therefore, $T(N, e)$ is still dominated by the term e. Setting $e = T(N, N) = N/lnN$, we see that the threshold in (11) holds well. Intuitively, because the frequencies of words drop off precipitously due to Zipf's Law, most of the computational complexity will be allocated to the top half of the lexical items anyway, such that a few exceptions located in the bottom half hardly make any difference. Finally, and most interestingly, it seems that the *psychological* frequency of words used by human language learners is strikingly detached from the empirical frequency of words, which raises some unsettling (and unanswered) questions — not just for the Tolerance Principle but for the computational system for linguistic information processing. I will return to these matters at the end of this chapter.

3.5 Remarks

The remaining chapters explore the utility of the Tolerance Principle in a large number of empirical cases. But first, a number of conceptual and methodological remarks are in order.

3.5.1 Smaller Is Better

An interesting consequence of the Tolerance Principle is that the number of exceptions must be relatively low to guarantee the productivity of a rule. Table 3.1 gives some values of N and the corresponding critical values of $\theta_N = N/\ln N$, which is a small proportion of N. The low level of tolerance for productivity may strike readers as absurd: surely a rule with more followers than dissenters — more than half, perhaps? — would be good enough. Needless to say, much of what follows is to provide evidence for the counterintuitive calculus of rules and exceptions. And for readers concerned with the existence of minority default rules such as the *-s* suffix in German noun plurals just described, an extensive discussion can be found in section 4.4.

Note that the proportion of tolerable exceptions drops quite sharply as N increases. In other words, "smaller" rules, those defined over relatively few items, can tolerate a relatively higher number of exceptions. Large rules, whose structural description includes many potential members, are more vulnerable. We have already seen hints of this peculiar property with respect to the statistical and structural aspects of English stress assignment (more in section 4.2). The sublinear growth of θ_N as a function of N is the most surprising, and in my view the most important, quantitative aspect of the Tolerance Principle. It suggests that, all things being equal, a learner that has a smaller vocabulary with respect to a rule will have a *better* chance of learning the rule than a learner with larger vocabulary. That is, if a learner somehow were to acquire all the relevant lexical items that may participate in a rule, a perfectly productive rule

3.5 Remarks

Table 3.1
The tolerance threshold for rules of varying sizes

N	θ_N	%
10	4	40.0
20	7	35.0
50	13	26.0
100	23	23.0
200	38	19.0
500	80	16.0
1,000	145	14.5
5,000	587	11.7

may not be learnable. By implication, a young child whose vocabulary may be limited has a better chance of learning the rules of language than an adult who generally knows more words. Throughout this project, I will explore, and exploit, the advantage conferred by the Tolerance Principle to a small vocabulary. It provides tantalizing clues to the age-old puzzle that children are superior learners of language despite — or perhaps *because of*— all the limitations of their cognitive system still in development.

3.5.2 Types, Tokens, and Artificial Languages

The earlier discussion of language processing places a good deal of emphasis on the token frequency of words and their constituents. Word frequency, approximated by Zipf's law, also figures into the calculation of expected time complexity in the derivation of the Tolerance Principle. The analytical solution (11), however, only makes reference to the *type* frequencies of N and e, the yeas and nays, in the end. This conception of productivity is in line with most previous theorizing from both generative and nongenerative perspectives (e.g., Baayen and Renouf 1996; Bybee 1995; Pierrehumbert 2003; Plunkett and Marchman 1991), and it is a fairly straightforward point. No matter how many times the child hears a single verb type that takes the *-d* suffix (e.g., *talked*), it would be folly to draw the sweeping conclusion that "*-d*" can attach to *all* verbs. A productive rule must be supported by a sufficiently large number of distinct types such that its open-endedness can be justified. Thus, rules do not become productive overnight and can only emerge after the accumulation of evidence. In section 4.1, I compare the development of verb past tense and noun plurals in English-learning children, showing that it is the type, rather than token, frequency that plays a decisive role in the acquisition of productivity.

While the current study largely focuses on the quantitative analysis of productivity in linguistic corpora, similar investigations can be carried in a laboratory setting using artificial languages to see how human subjects form rules and generalizations. In a series of papers (e.g., Hudson Kam and Newport 2005, 2009), Newport and colleagues have explored how token frequency affects the learning of rules. They find that children tend to aggressively pursue the statistically dominant pattern of variation at the expense of others, whereas adults appear to match the probabilities of variants in the input. The difference between the artificial language studies and the Tolerance Principle is exactly at the level of tokens vs. types. In Hudson Kam and Newport 2005 and similar studies, the usage of an individual item is inconsistent — for instance, a noun may be probabilistically used with multiple determiners or may form the plural with multiple endings. The closest analogy in language would be the case of morphological doublets (e.g., *dived* and *dove*) and dative alternations (e.g., *I give John a book* vs. *I gave a book to you*) where more than one form coexist.[11]

By contrast, the Tolerance Principle concerns cases where some items take one form while other items, a distinct set, take another form, thereby creating tension between rules and exceptions. It is possible to design studies that mimic the problem of productivity in acquisition. Indeed, experimental work with Kathryn Schuler and Elissa Newport using the artificial language paradigm has produced near-categorical support for the numerical predictions of the Tolerance Principle (Schuler, Yang, and Newport 2016). Children between the age of 5 and 7 were presented with nine novel objects with labels. The experimenter produced both the "singular" and the "plural" form for each noun as determined by its quantity on a computer screen. In one condition, five of

---

11 On the topic of doublets: the Tolerance Principle is likely applicable although there are some uncertainties about how to carry it out. Suppose that the learner encounters both *dived* and *dove* in the input data. Suppose further that, at least for present purposes, these two forms are functionally equivalent and completely interchangeable (as opposed to having been differentiated in some sense as suggested by Clark 1987 and others). With respect to the "-*d*" rule, then, the verb *dive* is both irregular and regular: How should we represent the dual status in terms of N and e in the Tolerance Principle? One possibility is that *dive* counts twice, one toward the exception (*dove*) and the other toward the rule (*dived*). A promising and better-motivated possibility, suggested by Constantine Lignos (2013) in his treatment of the postnasal plosive deletion in the history of English phonology, is that the learner will only use the more frequent of the variants in productivity consideration, similar to what Newport and her colleagues' research suggests. These issues will have to be left for future investigation.

the nouns share a plural suffix and the other four have individually specific suffixes. In another condition, only three share a suffix and the other six are all individually specific. Thus, the nouns that share the suffix are the regulars and the rest of the nouns are the irregulars. The choice of 5/4 and 3/6 was by design: the Tolerance Principle predicts the productive extension of the shared suffix in the 5/4 condition because four exceptions are below the threshold ($\theta_9 = 4.1$), but no generalization in the 3/6 conditions. In the latter case, despite the statistical dominance of the shared suffix as the most frequent suffix, the six exceptions exceed the threshold. When presented on additional novel items in a Wug-like test, nearly all children in the 5/4 condition generalized the shared suffix on 100% of the test items in a process akin to the productive use of English -*ed*. In the 3/6 condition, almost no child showed systematic usage of any suffix, much like speakers trapped in a paradigmatic gap (Baerman and Corbett 2010; Halle 1973; see chapter 5). The results from adult learners, however, were completely different. In both conditions, subjects matched the token frequencies of the suffixes, favoring the suffix that was the statistical majority.

In the closing part of this chapter, I will return to Schuler et al. (2016)'s results and their implications on how children make use of frequency information in language learning. The issue of type- and token-based will be picked up again in chapter 7, with some speculation on why children and adults produce different learning outcomes in various studies.

3.5.3 Effective Vocabulary and Variation

The Tolerance Principle allows room for variation in the transient stages of language acquisition as well as in the stable grammars of individual speakers. By hypothesis, productivity is determined by two integer values (N and e), which are obviously matters of individual vocabulary variation. For instance, many processes in English derivational morphology are inherited from Latinate words, which are quite rare in the primary linguistic data or even conversational English; see chapters 4 and 6. Consequently, derivational morphology only gets fully formed by school age, after children are exposed to a more learned vocabulary (Jarmulowicz 2002; Tyler and Nagy 1989). Inflectional morphology, by contrast, is learned very early (Brown 1973), presumably because the learner has access to higher volumes of data — both tokens and types — from which the inflectional rules can be formed. In chapter 4, I trace the development of both inflectional and derivational rules of English.

The cumulative process in which children acquire their vocabulary can lead to interesting dynamics in productivity. Generally speaking, words with higher token frequencies tend to be acquired earlier. A young child is likely to determine the productivity of the rules from a relatively small vocabulary of high frequency items. As we will see in chapter 4, children's acquisition of morphology and phonology is significantly affected by the frequency of words that enter into the tabulation of productivity, and children with different rates of vocabulary acquisition may also show significantly different developmental trajectories.

In addition, the productivity of a rule may change during the course of language acquisition. For instance, suppose that a rule is applicable to $N = 50$ items, of which $e = 20$ are exceptions. The Tolerance Principle predicts that the rule cannot be productive since the critical threshold for exceptions is at $\theta_{50} = 13$, which is fewer than 20. However, it is possible that at some initial stage of learning, the child has only acquired a subset of the lexical items, with the values $N' = 20$ and $e' = 5$. Such a mixture does warrant productivity ($5 < \theta_{20} = 7$), a conclusion that is likely overturned as the child acquires more words. Analogously, an ultimately productive rule (e.g., $N = 50, e = 10 < \theta_{50} = 13$) may be represented by an early and very partial sample ($N' = 15, e' = 8$), which will tip the balance against productivity ($8 > \theta_{15} = 6$). The learner would in fact lexicalize all fifteen items, only to reverse course and discover the broad applicability of the rule when her vocabulary becomes sufficiently representative. These cases are not hypothetical: an initially productive rule that loses out later can be found in the acquisition of English stress (section 4.2) and the initially lexical rule that gains productivity over time is the regular past tense "-*d*" (section 4.1).

Because the Tolerance Principle lives and dies by the number, a few remarks on our quantitative methodology are necessary. In light of the effective vocabulary and individual variation discussed above, the most faithful execution of the Tolerance Principle should be based on an individual learner/speaker's vocabulary. This is clearly very difficult, if not impossible, to follow in practice. For most studies, then, I provide two remedies to sidestep these challenges. First, I generally focus on rules and generalizations that are uniformly acquired by all individuals in the terminal state of language learning. For example, every English speaker eventually learns the "-*d*" rule for past tense and the double-object construction for dative verbs. Additionally, psycholinguistic studies show that children acquire these rules and generalizations at a fairly

3.5 Remarks

young age, which again suggests their uniformity throughout the relevant population. Second, I strive to obtain lexical statistics that are most appropriate for the case at hand. When the topic concerns child language acquisition, I primarily draw from child-directed language data in the public domain (CHILDES; MacWhinney 2000). For the acquisition studies of English, I have collected five million words of child-directed North American English from CHILDES, which at the time of writing represents the totality of English data in the public domain; it roughly corresponds to about a year of linguistic input from many English-learning children (Hart and Risley 1995). Thus, the corpus should be regarded as the input to a "typical" English-learning child, the type of linguistic information that most children will probably be exposed to during the years of language acquisition. I will make various adjustments to the child-directed corpus to mimic the developmental stages in acquisition, as chapter 4 explains in detail. When child-directed data is not available, I can only turn to other types of lexical databases. As a guiding principle, I typically include only words that appear at least once per million as estimated from large-scale corpora. I do so in part because the corpus data from various languages differ considerably in quantity, which makes an absolute frequency threshold impossible. Also, words that appear at least once per million can be available to most language users (Nagy and Anderson 1984), so I avoid corner cases caused by rare or obscure vocabulary items.

3.5.4 Recursive Tolerance and Structured Rules

The possibility of using the Tolerance Principle in a recursive fashion is perhaps the most interesting and subtle aspect of its application to language acquisition.

The formulation of the Tolerance Principle assumes that the exceptions to a rule form a monolithic list: the exceptions form a plain list sorted by frequency. This is clearly the case of English irregulars in relation to the "-d" rule because none of the irregular classes is productive. But exceptions needn't be deprived of regularities within. It is possible for language to make use of "nested" rules like a Russian doll. Consider an abstract example in (12).

(12) R_1: If $[+A, +B]$ THEN X
  R_2: If $[+A]$ THEN Y
  R_3: Z

where A and B are structural features of the lexical items under consideration. The Elsewhere Condition asserts that if an item has the feature $[+A, +B]$, then R_1 applies ("do X"). If an item has the feature $[+A]$ (but no $+B$), then R_2 applies ("do Y"). And R_3 serves as the default for items that do not meet the structural description of either R_1 or R_2: "do Z." Thus all three rules are productive as long as language learners are attuned to the feature specification of the lexical items. But the productivity of the rules may not be reflected in their type frequencies. For instance, suppose the hypothetical language in (12) has 90 words with the feature $+A$ out of which 70 also have the feature $+B$, but only 10 words that have neither $+A$ nor $+B$. In other words, R_1, R_2 and R_3 account for 70, 20, and 10 words respectively, but these numerical differences do not mean that some rules are "more" productive than others. In fact, the failure to incorporate the structural properties of morphological classes is one of the most persistent criticisms of purely quantitative measures of productivity (see Bauer 2001; Bolozky 1999; Van Marle 1992).

If the rules are ordered by the Elsewhere Condition as in (12), an item with the feature $[+A, +B]$ gets directly shipped off to R_1 and will be of no concern for rules ordered below it. Thus, none of the 70 items that go with R_1 would constitute exceptions for R_2 or R_3. But it is possible that child learners do not arrive at the organization of rules in (12) instantly. For example, they may initially fail to recognize the features on the lexical items and attempt to treat R_3 as the "global" default and try to accommodate the items that follow R_1 and R_2 as exceptions.

The numerical values of our hypothetical example will not favor R_3, the structurally more general rule, or any rule for that matter. Since there are 100 items in total, a productive rule defined over the *entire* set of items cannot tolerate more than $\theta_{100} = 22$ exceptions — even R_1 has a fatal 30 — which would force learners to lexicalize everything. This is clearly not a palatable state of affairs; as a linguist might say, it misses important *structural* generalizations.

In the spirit of the Tolerance Principle as an evaluation metric, we suggest the following general learning strategy:

(13) *Maximize Productivity*
  Pursue rules that maximize productivity.

Language favors rules that work. Strategy (13) embodies the Tolerance Principle as an evaluation metric that guides learners to discover productive rules (Figure 1.1). The failure to find productivity over a lexical set N provides the

3.5 Remarks

cue for learners to consider different rules, especially rules that divide the lexical items into subsets.

If child learners are attuned to the feature specification in our hypothetical example (12)—perhaps they *become* attuned because no productive rule will emerge without doing so—then any reasonable inductive model is able to detect the structural generalizations (i.e., the three rules). The Tolerance Principle can be applied recursively and succeeds in all cases. It is of course possible that the three rules have exceptions of their own. For instance, there may be $[+A, +B]$ words that do not follow X and are thus exceptions to R_1. In the example here, of course, there cannot be more than $\theta_{70} = 16$ such words if R_1 is to maintain its productivity.

The subdivision of the lexical N can lead to other rule-organization strategies. Consider another hypothetical example:

(14) R_1: If $[+A]$ then X
         R_2: If $[+B]$ then Y
         R_3: Z

Again, assume that there are 70 $[+A]$ items, 20 $[+B]$ items, and 10 unmarked items that follow the respective rules in (14). And again, none of the rules is sufficiently large to tolerate the others as exceptions. Here the principle of Maximize Productivity can also guide learners to discover the regularities in (14), when no productive rule emerges over the items without attending to their feature specifications.

Schematically, the examples in (12) and (14) can be represented as in Figure 3.5. We note that the nested rules are quite similar to the notion of levels in Lexical Morphology (Kiparsky 1982; Pesetsky 1977; Siegel 1974) where certain rules must apply before others as well as to the treelike hierarchical structures in Network Morphology (Brown and Hippisley 2012; Corbett and Fraser 1993), to characterize the generality and specificity of morphological rules. The Tolerance Principle can be regarded as a motivation for postulating theoretical analyses in these frameworks.

The recursive application of the Tolerance Principle, I believe, is critical for the learner to acquire the complex structures in the world's languages. The complete dichotomy in the English past tense system—a productive rule and a lexicalized list of exceptions—appears to be rare in a broad crosslinguistic context. An example of nested rules in (12) and Figure 3.5(a) can be found in the English nominal suffixes that attach to adjectives (Section 4.3). According to most descriptions of these suffixes (e.g., Plag 2003), *-ness* is the most

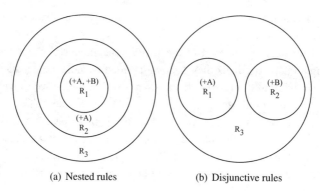

Figure 3.5
Recursive applications of the Tolerance Principle to detect structured rules

general rule that has no obvious restriction on the adjective stem, while *-ity* is nested within, because it productively attaches to adjectives that end in *-ible*, *-ic*, and *-al*. As such, adjectives with these specific phonological properties are handled by *-ity* and do not constitute exceptions to the *-ness* suffix. In computational terms, a word like *credible* is shipped to the *-ity* "subroutine" without ever coming across the *-ness* rule. These subregularities can only be revealed when learners fail to detect a productive rule over the entire set of adjectives: as we will see in section 4.3, none of the English nominalization suffixes come close to the requisite quantity to tolerate the others as the exceptions. The Principle of Maximize Productivity prompts learners to seek productive rules within the subdivisions of the adjectives — the Tolerance Principle will be applied recursively.

The disjunctive rules in (14) and Figure 3.5(b) can be observed in the metrical stress system of English (section 4.2). I show that the failure to detect a productive rule for stress assignment over the entire vocabulary leads to the subdivision into nouns and verbs, within which productive rules *can* be found. Similarly, the German plural system (section 4.4), can have a productive rule ("add *-s*") that applies to very few nouns because morphosyntactic and phonological features help partition nouns into several classes, all of which contain productive rules within, and thus do not constitute exceptions to the *-s* suffix.

The recursive application of the Tolerance Principle halts when productive rules are identified. Or, in the limiting case, there is no plausible way to discover productive rules. In chapter 5, I argue that the failure to detect productive rules is exactly where paradigmatic gaps arise in the synchronic grammar, and

3.5 Remarks 75

the changes in the productivity of rules, which may be predictable on a numerical basis, can help us understand how language change gets started. All the same, learners must be sensitive to the attributes over which the data is subdivided; this brings us to the next remark on the Tolerance Principle.

3.5.5 Structures and Statistics

A metric of productivity is only as good as the set of lexical items it evaluates. No Tolerance calculation can get off the ground without a precise specification of N and e: What makes a rule? How to count the exceptions?

In linguistic analyses, theorists are free to draw whatever generalizations from the data as they sees fit. Likewise, a statistical model can be applied to identify significant patterns in a sample of language use. Ultimately, however, we are concerned with the rules and generalizations that are cognitively accessible to language learners and form part of their I-language (Chomsky 1986). The distinction between what's logically possible and what's empirically meaningful is evident in our discussion of inductive learning in section 3.1. The capacities and limitations of the Yip-Sussman model are entirely determined by the "primitives" crafted by the designer, which define the type of rules that the model can in principle derive. Similarly, every time we apply the Tolerance Principle, we are committing to a specific analysis of the data—along with the values of N and e—which amounts to a claim about the psychological system of language and learning.

Here we are in somewhat uncharted territory. Despite several decades of empirical work, we are nowhere near a complete theory of language acquisition. In this sense, all subsequent discussion of productivity and learning will be tentative because we cannot be sure, at least in the general case, whether we are dealing with the cognitive hypotheses that child learners may actually entertain, or mere descriptions that analysts have dreamed up. Fortunately, these stumbling blocks can be partially overcome. On the one hand, the study of language development provides unambiguous cues for the learners' grammar. For instance, when children produce mistakes that deviate from adult language, such as the overgeneralization of rules (*hold-holded* in the past tense and *I said him no* in the double-object construction), they must be exercising productive options in their I-language because these forms are not available in the input. Furthermore, we have accumulated considerable evidence on the type of linguistic features and regularities that children selectively attend to. For example, in word segmentation tasks, 9.5-month-old infants prefer stressed syllables as cues for word boundaries over statistical information (Johnson and Jusczyk 2001), which helps us understand the acquisition of metrical stress (section 4.2). Similarly, a study of morphosyntactic acquisition (Gagliardi and Lidz 2014) shows that children make use of the phonological properties of words to determine noun-gender classes even though the semantic cues are statistically equally strong — which helps narrow down the range of the learner's inductive choices (see also Melançon and Shi 2015; Mills 1986; Mulford 1985; Pérez-Pereira 1991). These findings can help us pinpoint, and evaluate, the structural generalizations in child language even though we may not quite understand why certain linguistic features are privileged over others. As a practical strategy, the empirical case studies presented in this work focus on phenomena that have been extensively studied by previous scholars; whenever possible, I also provide independent evidence, from language acquisition, processing, and change, to establish the rules and regularities subject to productivity calculation.

*

A confession is in order before we plunge into rules and numbers. It has to do with the unreasonable effectiveness of the Tolerance Principle, which will come through in later chapters. The derivation of the tolerance threshold relies on several assumptions that, in most empirical cases, are at best only approximately true. Specifically, the term $\theta_N = N/\ln N$ presupposes that (a) any N words in a linguistic sample follows Zipf's law, and (b) the harmonic number $H_N = \sum_{i=1}^{N} 1/i$, which figures in the estimation of word probabilities, is approximated by $\ln N$. But we can easily verify that (a) and (b) are only true when N is very large (i.e., when we are dealing with very large linguistic corpora). Most of our empirical studies, however, are concerned with N's that are quite small, ranging from a dozen to a few hundred; after all, we are dealing with a young language learner's vocabulary, which will not follow Zipf's law very faithfully. For each case study, it would have been possible to estimate the empirical frequencies of words and solve for the cost of rules and exceptions — that is, $T(N,N)$ vs. $T(N,e)$ — numerically. I have not done so because getting reliable frequency estimates for a young learner's effective vocabulary is close to impossible; it also would have been hideous and no fun.

Or perhaps even completely futile. I now believe that during the courses of rule learning, the empirical frequencies of words are ignored *entirely* and children only keep track of the effectiveness of a rule (i.e., N, e, and θ_N) and

3.5 Remarks

nothing more. Recall the artificial language study by Schuler, Yang, and Newport (2016) reviewed earlier. Recall that children learned nine nouns, some of which share a plural suffix while others have idiosyncratic suffixes. By design, and following the Tolerance Principle, the nouns appeared in the input with frequencies approximately following Zipf's law. In both conditions (5/4 and 3/6), the nouns with the shared suffix were the most frequent items. We did so in order to ensure that children did in fact learn the nouns and their suffix; a post-test evaluation showed that as planned, even the lower-frequency nouns were recognized by children with greater-than-chance probabilities. Our study found that children generalized the suffix in the 5/4 condition but not in the 3/6 condition, just as predicted by the Tolerance Principle. This is striking because it is in fact *more* efficient to list everything if the empirical frequencies of the words were used. In the 5/4 condition, for instance, full listing would place the five most frequent nouns with the shared suffix in the position of one to five, the maximally efficient arrangement. To have a productive rule means to list the four least frequent items first, forcing the more frequent, rule-following items to wait, which clearly results in higher processing cost — if complexity were calculated with empirical frequencies. But as the experimental results show, children categorically postulated a productive rule.

I am at a loss as to why the Tolerance Principle appears to work so well to account for language learning. It is as if the assumptions that underly its derivation, which are meant to approximate reality, in fact *are* the reality. Here is an educated guess. Due to the serial nature of the mental lexicon or possibly semantic memory more generally (e.g., Sternberg 1969; Tulving 1972), words are inevitably placed on a sequential list (Murray and Forster 2004). This would immediately explain why it's not word frequency but the logarithm of word frequency, which approaches the linear function of rank most closely, that correlates with lexical decision latencies (Howes and Solomon 1951). The positions of words on a list may be used as a proxy for word frequencies in the external world, which the mind may not be capable of tracking precisely. That is, Zipf's law is an *exact* description of the mental reality of words, and is only an *approximate* description of their empirical manifestations — and hence works better only on very large linguistic corpora. If this conjecture is correct, we would be able to understand why the Tolerance Principle is an effective calculus for the psychological price of productivity.

These uncertainties aside, let's forge ahead. The proof of the pudding is in the eating, even though it's not clear how the pudding is made. The following chapters study a broad range of problems for which the Tolerance Principle provides a unified solution. To the extent that they are successful, these studies also amount to a defense of the serial nature of the Elsewhere Condition as a deep theoretical principle that has independent motivation from language processing and acquisition (Anderson 1969; Aronoff 1976; Brown and Hippisley 2012; Chomsky and Halle 1968; Kiparsky 1973; Stump 2001), as opposed to alternative formulations in Optimality-theoretic approaches (Bakovic 2013; Prince and Smolensky 2004). Chapter 4 establishes a range of "positive" examples: when there are productive patterns embedded in language, the Tolerance Principle can help ferret them out in the face of exceptions. Chapter 5 looks at the other side of the same coin: sometimes exceptions can overwhelm and productivity will break down—which doesn't take a lot, as can be seen in Table 3.1. I present a series of numerical results to show that when productivity fails, the learner needs to lexicalize everything that results in defective morphology and triggers linguistic change. Finally, chapter 6 turns to the theory and acquisition of syntax. I suggest that the Tolerance Principle can help children find the appropriate scope of inductive generalizations without succumbing to the familiar trappings of indirect negative evidence. I introduce the notion of *sufficient positive evidence*, which is derived from the Tolerance Principle, to resolve the puzzling case of English dative constructions (Baker 1979) and other problems.

Now we have the hammer, let's find some nails.

4 Signal and Noise

This chapter calibrates the machinery of the Tolerance Principle. It runs the gamut from morphology and phonology, featuring many prominent problems in linguistic theory and language acquisition.

It must be said that the execution of the Tolerance Principle follows a tedious mechanical routine:

(1)   a. Obtain a rule R along with its structural description and structural change.
      b. Count N, the number of lexical items that meet the structural description of R.
      c. Count e, the subset of N that are exceptions to R.
      d. Compare e and the critical threshold $\theta_N = N/\ln N$ to determine productivity.

I contend that a mechanical model is exactly what's required of a rigorous approach to language. For example, in the study of the agentive or instrumental nominalization suffix -er, we encounter words such as *player*, the expected use of the suffix on the stem *play*, as well as pseudo suffixed words such as *letter* and *rubber*, which may be analyzed as the stem *let/rub* combined with -er (meaning "someone that lets/rubs"). This may strike readers as silly but as we will see, that's the same mechanism at play when English speakers attempt to analyze -er as a suffix in *brother* (e.g., Rastle, Davis, and New 2004). After all, *let* and *rub* are very common verbs and the meanings of "someone that lets/rubs" are completely legitimate; it's just the form of *letter* and *rubber* is already paired with other meanings that have nothing to do with *let* or *rub*. Likewise, *counter*, at least in its more common usage as a short form of *countertop*, also contains a spurious -er. If we are to reconstruct the mechanics of language learning, then the productive status of -er must be able to withstand the instances of misanalysis — see also *banner, hammer, summer*, etc. — as exceptions.

More concretely, we envision the learner experimenting and evaluating the grammatical hypotheses in an incremental fashion as the input data is processed. Each rule R can be identified with a tuple (N, e). The learner traverses through a sequence of grammars as learning proceeds:

(2) $G_1(N_1, e_1) \to G_2(N_2, e_2) \to G_3(N_3, e_3) \to ... \to G_T(N_T, e_T)$

Each transition $G_i \to G_{i+1}$ is enabled by the rejection of G_i that is determined to be unproductive by the application of the Tolerance Principle.

The learning scheme in (2), therefore, is transformational, which was — and still is, as discussed in chapter 7 — rejected as a suitable model for paramater setting in earlier work (Yang 2002). It belongs to the class of hypothesis-testing models in psychology (Levine 1975; Trabasso and Bower 1975) and more familiarly the error-driven learning paradigm in the study of language learnability (Berwick 1985; Chomsky 1965; Gibson and Wexler 1994; Wexler and Culicover 1980). In most formulations of hypothesis testing, the current grammar or rule is evaluated against a specific item. For example, in the triggering model of parameter setting (Gibson and Wexler 1994) and the constraint demotion model for Optimality-Theoretic ranking (Tesar and Smolensky 2001), an instance of the input data inconsistent with the current grammar leads learners to revise their grammar. The Tolerance Principle, by contrast, operates on a cumulative *set* of words — two numerical measures to be precise — that the learner has acquired so far. A productive rule may suddenly be rejected if an additional exception appears in the input; likewise, an unproductive rule may be revived if additional items come to its support. Language acquisition must hit a moving target, because the status of a grammar is established on the balance between rules and exceptions. This view of language learning has immediate implications for the modern grammar-based approach to language change initiated by Halle 1962, which we develop further in chapter 5.

Let's start with a very simple case that provides a clear illustration of the Tolerance Principle in action. Recall the German participle system, which has been extensively studied in both language acquisition and language processing (Clahsen 1999; Fleischhauer and Clahsen 2012; Marcus et al. 1995). Unlike the English past tense, which has myriad rules for suffixation, German past participles bifurcate into the regular *-t* and the irregular *-n* suffix.[1] As reviewed in section 2.3, German-learning children overregularize *-t* but almost never overextend *-n*. Furthermore, the irregularly suffixed forms are processed in a frequency sensitive fashion, which suggests some form of memorization. By contrast, the reaction time for the regulars is not differentiated by the frequencies of the inflected form, which provides evidence for a productive rule that applies across the board (Clahsen 1999). The German past participles, then, constitute a system of rule plus lexicalized exceptions par excellence.

---

1 There are a very small number of irregulars that also have unpredictable stem changes. These are excluded from the analysis, but including them would not change the outcome of productivity calculation.

The Tolerance Principle claims that the number of -*n* attached verbs must be sufficiently low to justify -*t* suffixation as the default rule. In the German CELEX database (Baayen, Piepenbrock, and Gulikers 1996), there are 1,700 monomorphemic verbs that take the -*t* suffix (Smolka, Zwitserlooch, and Rösler 2007). By contrast, there are only 200 irregular verbs that use -*n*. The rule of -*t* suffixation, then, would have the value of N at $1{,}700 + 200 = 1{,}900$, and e at 200, the exceptions that defy the rule. But the number of exceptions is well below the critical threshold at $\theta_{1900} = 252$: the status of -*t* as the default suffix can be predicted completely numerically.

What follows are over a dozen cases of similar quantitative analyses, ranging from inflectional to derivational morphology, from interlearner variation to intralanguage rules that operate on different sets of words. I show that productivity results from the tension between signal and noise, which learners will need to detect in an incremental fashion as they acquire the vocabulary.

4.1 When Felt Becomes Feeled

Let's begin our application of the Tolerance Principle with something very familiar, the acquisition of past tense in English.

My analysis draws on a five-million-word corpus of child-directed North American English (MacWhinney 2000). These utterances are tagged for part of speech and all past-tense tokens are then extracted.[2] In all there are 1,022 unique verbs that appeared in past tense, of which 127 are irregular. To be productive, the regular rule for the realization of PAST as -*d* must be able to withstand the numerical assault from the irregulars. For the values of $N = 1{,}022$ and $e = 127$, we see that the productivity of the "-*d*" rule still has considerable breathing room: the critical threshold of exceptions is $\theta_{1022} = 147$. This is reassuring: even young children can securely establish the productivity of the -*d* rule. These measures are in general agreement with results from other corpora. For instance, in the 452-million-word COBUILD Corpus (Sinclair 1987), 1,062 verbs appeared in the past tense at least once per million. Of these, there are 137 irregular verbs, which also fall below the critical threshold of 152 (or θ_{1062}): the "-*d*" rule is again justified.

But the more interesting question in past-tense acquisition is not about the terminal state of the language, where the status of the past-tense rules is not in

---

2 Unknown words are discarded. All quantitative analysis of English in the present work used GPosTTL, a statistically enhanced Brill tagger (1995) available at http://gposttl.sourceforge.net.

doubt. We would like to know when and how young children recognize the productivity of the rules during the course of language learning. The emergence of overregularization errors in child language (Marcus et al. 1992) suggests that productivity does not come overnight—Adam's first error, "What that feeled like," appeared just before his third birthday. Let's see how this process unfolds.

Suppose the learner has derived a set of morphological rules with some some suitable model of inductive learning. For concreteness, we consider the rules produced by Molar-Yip-Sussman model (2001) reviewed in chapter 3.[3]

(3) a. $ɪ → æ / \_\_ ŋ$ (*sing-sang, ring-rang, ...*)
$i → e / C \_\_ [d, t]$ (*feed-fed, lead-led, meet-met, ...*)
$[aɪ, ɔ, o] → u / [l, n, r] \_\_$ (*fly-flew, grow-grew, know-knew, draw-drew, ...*)

b. PAST → d

I focus on the irregular rules in (3a) because they appear to have a fighting chance to be productive. Most of the irregular rules learned by the model are like the "ought" rule ("rime → ɔt/ __ "), which has few structural restrictions because the verbs it does take are phonologically very diverse.[4] As such, they tend to have relatively large values of N. For instance, the "ought" rule says that any rime can change to /ɔt/, which of course would include every verb in English. But only a tiny number of verbs would follow the rule—six to be exact—with an enormous number of exceptions, relegating the rule to the dustbin of unproductivity right away. In contrast, the rules in (3a) are structurally fairly restrictive and have corresponding low values of N, which increases the odds for productivity; recall the numerical results in Table 3.1 showing that small values of N tolerate a proportionally larger number of exceptions. Furthermore, at least some of the rules in (3a) are known to facilitate very limited degrees of overirregularization. The "sing-sang" rule ($ɪ → æ / \_\_ ŋ$) is the only irregular pattern occasionally extended in children's speech (*bring-brang/brung*; Xu and Pinker 1995), and the other two rules also

---

[3] I thank Ray Molnar for making his code available. For clarity of presentation, I use the IPA symbols instead of the distinctive features as implemented in the model. To take an example, for the class that contains *fly, draw*, and *know*, the model identifies the prevocalic consonant as having the features [+sonorant, +voiced, +coronal, +anterior]. The only segments in English that meet these criteria are in fact /l/, /r/, and /n/.

[4] A few other rules cover only one verb (e.g., *say-said, come-came*) in the child-directed corpus; they are in effect suppletive and have no potential for generalization at all.

4.1 When Felt Becomes Feeled

seem to favor irregularization (e.g., *gleed-gled*) in studies of adult past-tense formation (e.g., Albright and Hayes 2003; Bybee and Moder 1983).

Of course, terms such as regular and irregular are hindsights, labels assigned by theorists on the basis of linguistic analysis: they need to be discovered by the child from the naturally occurring data of English past tense formation. To model the development of the past tense, I evaluate the productivity of the rules in (3) by gradually enlarging the inventory of verbs, so as to mimic the cumulative effect of vocabulary growth over time. Note that a verb being merely available to the learner is quite different from the learner having learned it. Ultimately the calculation of productivity rests on the effective vocabulary that the child has successfully internalized (section 3.5.3), which is notoriously difficult to estimate (Dale and Fenson 1996; Huttenlocher et al. 1991). In section 4.1.2, I consider the acquisition of the past tense by several individual learners for whom there is a reasonable amount of acquisition data. However, we can only consider the learning trajectory of a "typical" English child learner. To the extent that the child-directed English corpus gives a reasonable approximation of the distribution of English words, a quantitative assessment of the rules will be informative.

The results are summarized in Table 4.1, where I use the format (N,e) to denote the number of verbs fitting the structural description of a rule and the number of exceptions to that rule. The rows represent verbs ranked by frequency. For instance, 100 (row 1) provides a list of the 100 most frequent verbs that appear in the past tense: all of them, of course, fit the rule "PAST $\rightarrow$ d" but 54 are irregulars, or $(100, 54)$. The productivity threshold for the regular rule is $\theta_{100} = 22$, which means that the "PAST $\rightarrow$ d" rule cannot be regular for a typical child learner with only 100 most frequent verbs. As we enlarge the number of verbs, Table 4.1 provides an approximation of English-learning children's vocabulary as their language develops, where smaller sets of words represent earlier stages of acquisition. The rules are evaluated for productivity for each stage.

It is instructive to observe how the productivity of rules changes over the course of learning.

4.1.1 Evaluating Irregulars

Consider first a very small verbal vocabulary where we only include the 100 most frequent past tense forms. As shown in the first row of Table 4.1, it is very difficult for any rule to be productive because the range of attested forms

Table 4.1
Productivity of English past-tense rules as a function of vocabulary size. Boldface denotes productive status, and "—" indicates that the rule is unavailable from the sample or trivially unproductive (i.e., assumes a singleton member). The three irregular rules in (3a) are represented by an example (*sing-sang*, *feed-fed*, and *fly-flew*).

Top N	sing→sang	feed→fed	fly→flew	-d	θ_N
100	—	—	**(8, 3)**	(100, 54)	22
200	**(3, 1)**	—	(10, 5)	(200, 76)	37
300	**(3, 1)**	—	(13, 8)	(300, 92)	52
500	**(5, 2)**	**(6, 3)**	(15, 10)	(500, 103)	80
800	**(8, 5)**	(11, 7)	(18, 13)	(800, 121)	119
1022	**(8, 5)**	(13, 9)	(22, 16)	**(1022, 127)**	147

is too diffuse. For instance, more than half of the 100 most frequent verbs are irregulars: the "-*d*" rule cannot be productive because the tolerance threshold for θ_{100} is only 22.

However, the "fly-flew" rule ([aɪ, ɔ, o] → u/ [l, n, r]) is more promising. Among the top 100 verbs, there are 8 that fit its structural description, of which 5 (in boldface) are inflected as prescribed:

(4) **blow**, cry, **fly**, **grow**, **know**, lie, **throw**, try

The 3 exceptions are in fact manageable ($\theta_8 = 3$) and the rule can be expected to be productive, even though no attested overgeneralization error can be found (e.g., *lie-lew*; Xu and Pinker 1995). But the productivity period of this irregular pattern would be very brief, and perhaps ends even before children start producing the past tense in speech. The top most frequent verbs include 11 candidates for the "fly-flew" rule.

(5) **blow**, cry, **fly**, follow, **grow**, **know**, lie, snow, swallow, **throw**, try

Now the exceptions have risen to 5, exceeding the tolerance threshold ($\theta_{11} = 4$). The rule never recovers as the vocabulary size increases, as can be seen in Table 4.1. It will become completely lexicalized from this point on.

The "feed-fed" rule is also likely to be transiently productive before dropping out of contention entirely. There are only a few verbs, irregular or regular, that fit the structural description. When the vocabulary size is small (e.g., in the top 300 verbs), there are only 3 verbs fitting the description (*feed-fed*, *meet-met*, *need-needed*) that make an appearance. It's unclear if any learning

4.1 When Felt Becomes Feeled

model can discover the structural regularity (i → e / C __ [d, t]), and the evaluation of productivity is a moot point. With the top 500 verbs, the rule probably becomes viable: 3 exceptions out of 6 are tolerable:

(6) cheat, **feed**, greet, **meet**, need, **plead**

Once again, its productivity cannot last. With the top 800 verbs, there are already more exceptions than rule-following items:

(7) **bleed**, cheat, **feed**, greet, heat, **meet**, need, **plead**, precede, repeat, treat

The rule will remain irregular from now on, and the boldface verbs will be lexicalized accordingly.

But another rule enters into the foray. The top 200 verbs introduce the "sing-sang" rule (ɪ → æ / __ ŋ):

(8) bring, **ring**, **sing**

which works for 2 out of the 3 verbs that fit its structural description. In fact, the rule may enjoy an extended period of productivity. Among the top 500 verbs, the 2 exceptions (*bring* and *swing*) cause no difficulty for the remaining 3:

(9) bring, **ring**, **sing**, **spring**, swing

The sustained productivity of ɪ → æ / __ ŋ, I believe, is the probable explanation for the only systematic "weird past tense error" observed in child English (Xu and Pinker 1995). It is consistent with Berko's 1958 irregularization test: only one out of the 86 children used an irregular form for novel verbs, and that form happens to follow the "ɪ → æ / __ ŋ" rule (*bang* and *glang*). As reviewed in Section 2.2, the "sing-sang" pattern accounts for 15 of the 40 overirregularization errors (out of 20,000 past-tense forms). The verb *bring*, in particular, is quite frequently overirregularized. In the entire collection of North American child English transcribed in the CHILDES database, the past tense of *bring* is used correctly (*brought*) 95 times, regularized (*bringed*) 6 times, and irregularized 9 times (*brang* or *brung*). It is quite likely that these errors were produced during the transient stage of productivity for ɪ → æ / __ ŋ.

But this stage won't last either. When we expand the vocabulary to the top 800 verbs, the composition of the data changes and a new pattern emerges:

(10) bring, *fling*, **ring**, **sing**, **spring**, *sting*, *swing*, wing

There are now 8 verbs that end in /ɪŋ/ but their past-tense forms are scattered all over the place: the three in boldface change the vowel to /æ/, the three in italic change the vowel to /ʌ/, one is idiosyncratic (*bring*), and the last is regular (*wing*). All of them lose because none is numerically dominant enough to tolerate the rest.

It is unlikely that the /ɪŋ/-ending verbs will find any productive pattern, at least in the "standard" variety of American English. Consider all the verbs that appear in the 51-million-word American English SUBTLEX-US Corpus which can be regarded as a relatively representative lexicon of the spoken language (Brysbaert and New 2009). There are only 14 verb stems that end in /ɪŋ/; their past-tense forms are exhaustively listed below:

(11) a. "ought": bring (1)
 b. ɪ → æ / __ ŋ: sing, ring, spring (3)
 c. ɪ → ʌ / __ ŋ: swing, string, sting, fling, cling, sling, wring (7)
 d. Regular: wing, zing, ding (3)

None of these patterns here is numerically large enough to achieve productivity because the maximum number of exceptions is only $\theta_{14} = 5$. The ɪ → ʌ / __ ŋ pattern, while the statistical majority, is still insufficient, although it may have a longer shelf life if a speaker for some reason did not learn some of the seven verbs that are the exceptions. Again, recall that some children do produce *brung* in their past-tense production. Putting aside the task-specific complications of the Wug test, adults do sometimes generalize *bing* and *gling* along the irregular lines (Berko 1958; see also Bybee and Moder 1983). For most speakers, however, the transient productivity of ɪ → æ / __ ŋ (or ɪ → ʌ / __ ŋ) will eventually crumble. The irregular verbs in (11) will lexically go with the irregular rules; *wing*, *zing*, and *ding*, by virtue of not being on any list, will be picked up by the "-*d*" rule like all the other regular verbs.

The evaluation of the /ɪŋ/ verbs, in particular, shows clearly that the Tolerance Principle leaves room for individual and dialect variation (Herman and Herman 2014). If a learner happens to receive input data where an overwhelming majority — as determined by the numerical relationship between N and e — follows a vowel-change rule in (11), then that rule may survive into the stable grammar as productive and may assimilate other /ɪŋ/-ending verbs. All the same, if a learner consistently draws a very skewed sample of words that overrepresents the exceptions, then a supposedly productive rule may collapse

4.1 When Felt Becomes Feeled

and lose its productivity. These possibilities plant the seed for language variation and change (chapter 5).

4.1.2 Adam, Eve, and Abe

So far I have not said a word about the productive rule of "add -*d*." When the vocabulary is small, the "-*d*" rule doesn't stand a chance. The irregulars tend to be the most frequent forms in the past tense. As shown in Table 4.1, the top 100 verbs have a majority of irregulars (54). The ratio of irregulars shifts to a minority (76) at 200 verbs but still doubles the tolerance threshold of 37 (θ_{200}). The productivity of "-*d*" will have to wait until the child learns an overwhelming number of regulars. The rise of the regular past tense requires just a little more than 800 verbs — 823 to be exact, which includes 122 irregulars, matching the tolerance threshold exactly.

Of course, the child may not need over 800 verbs to learn the regular past. Our corpus results are based on the *input* to the learner: as discussed in section 3.5.3, children's productivity calculation depends on their effective vocabulary, which would be a particular subset of the input. When the appropriate data is available, the Tolerance Principle can provide quantitative predictions as to *when* productivity emerges in child language; this predictive approach to language development for individual learners is completely novel as far as I know.

Consider again, for the last time, the U-shaped learning curve in English past tense. Like previous authors (e.g., Marcus et al. 1992), I assume that the very first instance of overregularization unambiguously marks the emergence of productivity for the -*d* rule. Adam, for instance, produced the first token of overregularization at the age of 2;11 — *What dat feeled like?* — thereby entering the dipping segment of the U-shaped curve. As shown in Figure 4.1, Adam's irregular past tense was perfect prior to 2;11, after which the irregulars occasionally fell under the "-*d*" rule.

It must be at this point — 2;11, and certainly no later — that the -*d* became productive for Adam. By the Tolerance Principle, he must have learned enough verbs (N) such that the irregular verbs (e) in his vocabulary could be tolerated. To test this prediction, I estimated Adam's effective vocabulary as follows. I extracted all the verbal lemmas he produced between his earliest recording session (at 2;3) and the point (2;11) where *feeled* was produced. A verb is considered to be in Adam's lexicon if any of its inflectional variants appears in his speech. That is, if Adam produced *walking*, then the verb *walk* is included in

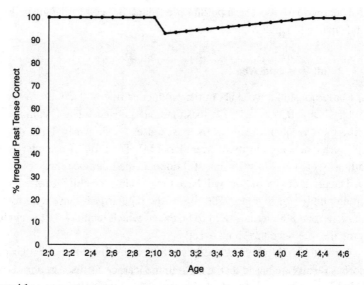

Figure 4.1
Schematic illustration of Adam's U-shaped past-tense development, adapted from Marcus et al. 1992. All irregular verbs were used perfectly until 2;11, when the first instance of overregularization took place.

his vocabulary even though the past-tense form *walked* may not have made an appearance. This method produced a list of 300 verbs, of which 57 are irregular. In order for the *-d* suffix to be productive, there should be more than θ_{300} or 53 irregulars. We are agonizingly close (53 vs. 57), and the difference can well be attributed to sampling effects. Since the CHILDES transcripts could not have recorded Adam's complete vocabulary, the regular verbs, which are in general less frequent than the irregulars, must have been undersampled to a greater extent. The more general point of this exercise is to highlight the critical condition for productivity. Only an overwhelming majority will suffice, and Adam rectified the "*-d*" rule only after the regulars thoroughly outnumbered the irregulars.

The quantitative considerations of the Tolerance Principle provide a concrete and rigorous approach to the problem of individual variation in language acquisition. That individuals in a linguistic community generally learn comparable grammars would follow if the composition of their vocabularies at the conclusion of acquisition is also comparable. That is, the N's and e's across individuals, which must vary to some extent, nevertheless reach the same decision regarding the productivity of grammatical rules. But children may acquire their vocabulary at different rates, along with different N's and e's, which can lead to differences in the calibration of productivity during the course of development.

Consider Abe, the son of the psychologist Stan Kuczaj (1976; 1977) and the great enigma in the study of English past tense. Abe's acquisition data has been in the public domain for a long time; over the years, he has become notorious for his extraordinarily high rate of overregularization. For instance, my own earlier work (Yang 2002) finds that the three children in Brown's 1973 Harvard project all did relatively well on past tense, with overregularization error rates of 1.8% (Adam), 7.8% (Eve), and 3.5% (Sarah); Abe, by contrast, had a whopping 24% of errors. The most detailed study of past-tense acquisition (Marcus et al. 1992) dismisses Abe's data as "chaotic". It certainly does not fit with the narrative that overregularization results from sporadic memory retrieval failures of the irregular form (Pinker 1995).

But there is nothing chaotic about Abe's past-tense acquisition. First, Abe contributed more past tense data than any other child in the public domain: the 24% error rate is based on 564 overregularized forms out of 2,350 opportunities and is clearly not an artifact of small sample size. Second, while Abe overregularized many verbs, he did show considerable improvement over time. In my reanalysis of his data (Yang 2002, 91), I found that for a set of problematic verbs that he overregularized very often initially (up to 50% to 70%), he showed considerable improvement over time, with the error rate down to about 5% two years later; see Maratsos 2000 for a similar observation. Finally, despite his extraordinarily high rate of overregularization overall, Abe did very well on a certain subset of verbs, precisely those for which *all* children do well. For instance, Abe, like other children, generally inflected the vowel-shortening verbs (e.g., *lose-lost, leave-left, shoot-shot* and *bite-bit*) correctly despite the relatively low lexical frequencies of these words in the input, which can be attributed to the collective effects of shared rules and general morphophonological processes (section 2.2). In sum, his past tense usage is qualitatively similar to that of other children; it is just that the "-*d*" rule somehow runs rampant.

It would be a mistake to assume that Abe's language-learning ability is somehow deficient. In my experience working with numerous samples of child language, he is among the most linguistically precocious children. Irregular past tense aside, his language development is by all measures well ahead of his peers. From the first recording session at the age of 2;3, Abe produced long and complex sentences such as *I want to make a book with monsters*. The legendary Eve, widely known in the language acquisition community for her remarkably early development of vocabulary and syntax (Brown 1973), had a Mean Length of Utterance (MLU) of 4.05 words at the age of 2;3 when she left the Harvard study. At this age, Abe had an impressive MLU of 4.53; Adam is far behind (3.09).

There may be many reasons for individual variation in children's language, including the amount, and the structural properties, of the linguistic input they receive (Yang, Ellman, and Legate 2015). But the simplest account for Abe's curiously high rate of past tense errors is that he's *too* good a language learner.[5] Abe is not only a prolific overregularizer but also a remarkably early one. His transcripts start at 2;3, and he was already overregularizing: *he falled*. Adam, by some distance the best irregular verb user, only started overregularization eight months later at 2;11. Abe, then, learned the productivity of the "*-d*" rule at least eight months before Adam, and possibly even earlier; unfortunately there are no transcripts before 2;3 so we cannot carry out Tolerance calculations as we did for Adam.

Eve is perhaps even more impressive. Her first overuse of *-d* came at 1;10 (*it falled in the briefcase*), a full year ahead of Adam. Extracting the verb stems from Eve's transcripts up to this point yields 163 distinct types altogether, including 49 irregulars, which is higher than the maximum threshold ($\theta_{163} = 32$). But there are strong reasons to believe that the undersampling of the regulars affected Eve's data much more than Adam's. From the beginning of the transcript to the first instance of overregularization, Adam's speech yields a token/type ratio of 19.24 (33,614/1,747); that is, on average, a new item appears in every nineteen words. For Eve, however, the token/type ratio is 11.8 (10,850/916): a new item every twelve words. Therefore, Eve's verb vocabulary is probably considerably larger than the attested 163; interestingly, her over-regularization rate is also much higher than Adam (7.8% vs. 1.8%). The token/type analysis also confirms that Abe has a larger vocabulary than Adam: for Abe to produce 1,747 unique words, he only needed just over 25,000 tokens (as opposed to Adam's 33,614).

---

5 Maratsos (2000, 202) reports that at age four, Abe scored 140 on the Peabody Verbal I.Q. test. Presumably Abe kept excelling at the verbal art; he is currently a partner at a corporate law firm in Austin, Texas (Abe Kuczaj, personal communication).

I cannot refine the estimate of children's vocabulary much beyond what is in the public domain. But it seems clear that the very different trajectories of *rule* acquisition may be caused by individual differences in *word* acquisition; after all, rules can only be learned on the basis of words. It is worth noting again that even in samples of child language that almost certainly underrepresent regular verbs, the number of irregular verbs still must be a very small fraction of children's vocabulary when the "-*d*" rule becomes productive: 19% for Adam and 18.6% for Eve. These findings support the counterintuitive prediction of the Tolerance Principle that rules must thoroughly overwhelm exceptions, as shown in Table 3.1.

4.1.3 Why Are Noun Plurals Easier to Learn?

The preceding discussion makes the case that morphological development is crucially dependent on the amount, and the composition, of input evidence the child receives. This ought to be a truism, although the role of the input has been downplayed if not completely ignored in the generative approach to language acquisition—not to its advantage, in my view. I return to these broader themes in chapter 7, but for the moment let's consider the acquisition of English noun plural-marking, which provides an instructive case study of how the input matters.

Since the pioneering work of Roger Brown (1973), it has been known that not all pieces of morphology are acquired at the same time. For English-learning children, a fairly consistent sequence of acquisition of inflectional morphology has been established, using Brown's well-known 90% usage criterion in obligatory contexts. The progressive -*ing* appears first, followed by noun plurals (-*s*), regular past tense (-*d*), and then third-person singular (-*s*). For lexical verbs, the progressive and third-person singular are exceptionless, and their order of acquisition is transparently determined by frequency: their frequencies of usage in the five-million-word child-directed English data are 73,292 and 23,779 respectively.

The comparison of noun-plural and past-tense acquisition reveals a more interesting pattern: the earlier acquisition of noun-plural marking, which generally takes place before the age of two (e.g., Mervis and Johnson 1991), is *not* a function of frequency. Quite the opposite; plural nouns appeared 69,246 times in our child-directed corpus, over 20% fewer than past tense (89,030). Here the Tolerance Principle helps to account for the discrepancy between the quantity of the input and the development of the grammar. While both the noun-plural and the verb past-tense systems have exceptions, their statistical compositions are very different. The irregular nouns are not nearly as frequent as the irregular verbs, and not nearly as top heavy in the frequency/rank spectrum. Of the top twenty most frequent plural-marked nouns in our child-directed English corpus, only four are irregular (*people, feet, teeth,* and *children*); for verb past tense, it's the complete reversal with only three being regular. Similar to the acquisition of past tense (Table 4.1), the balance between rule and exceptions for the noun-plural marker *-s* can be observed in Table 4.2.

Table 4.2
The productivity of the noun-plural *-s* is established early and consistently.

Top N	e	θ_N
10	2	4
20	4	7
50	4	13
100	6	22
200	9	38
400	13	67
1000	24	145

Table 4.2 shows that any sample of plural nouns, even very small ones that include only the very frequent English words, will decisively support the productivity of the *-s* suffix. Note that Brown's 90% criterion does not necessarily establish the productivity of the morphological rule. As usage-based theorists remind us (Tomasello 2000b), the child may have just memorized the plural forms in adult language and managed to retrieve them reliably; see section 2.1 for arguments and counterarguments. The unambiguous evidence for the acquisition of productivity is, once again, overregularization, for it could not have been retained from the input. Following this logic, we find that Adam's first instance of an overregularized noun (*peoples*) appeared at 2;8, or three months prior to his first instance of an overregularized verb (*feeled*). I extracted all the nouns he produced up to this point and manually removed the mass nouns and others that do not readily lend themselves to pluralization. There

are 431 nouns in all, of which 10 are irregulars, which clearly justfies the postulation of the -*s* suffix as the default.[6] As predicted by the Tolerance Principle, the -*s* suffix on nouns is easier to learn than the -*d* suffix on verbs not because it's more frequent in the input, but because it has fewer exceptions to overcome.

4.2 A Recursive Approach to Stress

The second case study in this chapter deals with the acquisition of English metrical stress. This is a long section because the stress system of English is highly complex and an acquisition study must build on the findings from speech perception, phonological development as well as morphosyntactic learning. I delve into this complicated problem because it provides a clear demonstration of how the Tolerance Principle serves as an evaluation metric to guide children in rule learning. Furthermore, the fracturing of the English stress system, with distinct stress rules for nouns and verbs, is an excellent example of the recursive use of the Tolerance Principle as observed in section 3.5.4. It is often the *failure* to discover a productive rule that leads learners to revise their grammatical hypotheses so that productivity can be subsequently reengineered. As we will see later in this work, the recursive use of the Tolerance Principle is crucial for understanding German noun plurals, English nominalization suffixes, paradigmatic gaps and language change (chapter 5), and dative constructions in English syntax (chapter 6). But first, I turn to the intricacies of English stress and the challenges it poses for child learners.

Why English stress? First, the stress system of English has played a significant role in the development of phonological theories (Chomsky and Halle 1968; Halle 1998; Halle and Vergnaud 1987; Hayes 1982, 1995; Liberman and Prince 1977) yet considerable disagreement remains: acquisition considerations may provide an independent platform on which these competing approaches can be evaluated. Second, there is now a reasonable body of developmental data on stress acquisition, both longitudinal and cross-sectional, and the major trends in the development of metrical stress have been identified. Third, and most interestingly, the metrical system of English is laden with

---

6 Eve's first instance of noun-plural overregularization (*feets*) appears in the same recording session as her first instance of overregularized verbs: the -*s* suffix for nouns became productive for Eve no later than the -*d* suffix for verbs.

exceptions, thanks to the extensive lexical borrowing in the history of the language. There has been no lack of theoretical apparatus for marking the exceptions — diacritics, extrametricality, lexically specific constraint ranking, etc. But the acquisition problem is fundamentally the same: words are not labeled as exceptions and the child needs to identify rules — and the exceptions to them — all at the same time.

4.2.1 A Sketch of English Stress

The English stress system is complex enough to have engendered a number of competing theoretical analyses, though several points of generalization are common to most. Roughly speaking, main stress in the nominal domain falls on a heavy penultimate syllable, and otherwise on the antepenult. In verbs, main stress falls one syllable closer to the word boundary: on a heavy final, and otherwise on the penult. Theories diverge only at a closer level of detail; here I consider only two formulations chosen largely for the explicitness of their mechanics.

Consider first Halle and Vergnaud's 1987 system (henceforth HV87). For nouns, the system is designed to capture the generalization that main stress falls: (i) on the final syllable if it contains a long vowel, (ii) on the penult if its rime is branching, and otherwise (iii) on the antepenult. Final syllables are considered extrametrical (Hayes 1982, 1995) if they contain a short vowel, which prevents final main stress. In verbs, stress generally falls on one of the final two syllables. Since antepenultimate stress is not at issue, extrametricality is not posited for verbs. Furthermore, the determination of quantity sensitivity is different for verbs, in that a word-final consonant is ignored. Otherwise, the system described above for nouns applies, yielding results wherein stress falls on a final superheavy syllable, and else on the penult. Operationally, the HV87 theory makes the following predictions for English stress:

(12) The HV87 (Halle and Vergnaud 1987) system
    a. Nouns
- If the final syllable contains a long vowel (VV), it receives primary stress (e.g., *kangaroo*).
- Otherwise if the penult is heavy (i.e., VV or VC^+, short vowel with at least one consonant coda), then the penult receives primary stress (e.g., *symposium, reluctance*).

- Otherwise the antepenult receives primary stress (e.g., *treachery*).

b. Verbs

- If the final syllable is super heavy (i.e., VV or VCC$^+$, a short vowel with at least two consonants in the coda), then the final syllable receives primary stress (e.g, *design*, *erupt*).
- Otherwise the penult receives primary stress (e.g., *abandon*).

Halle (1998; henceforth H98) departs from previous approaches in a number of respects. The adoption of Idsardi's (1992) foot construction system provides an alternative formulation of extrametricality: Halle's "edge marking" rules can ensure that for nouns, a final syllable containing a short vowel will not be part of the foot containing main stress. The machinery of H98 is very complex; for an exposition, see Legate and Yang 2013. Here I only summarize its predictions for the primary stress assignment of nouns; the net outcome of verb stress assignment is the same as the HV87 system.

(13) The H98 (Halle 1998) system

a. Nouns

- If the penult is heavy (i.e., VV or VC$^+$), then it receives primary stress (e.g., *horizon*, *tempest*).
- Otherwise the antepenult receives primary stress (e.g., *abdomen*, *hospital*).

b. Verbs

- Same as HV87 above (12b).

Several remarks are in order before we examine how a child acquires English stress assignment. First, despite the predictive similarities between the two theories, which are achieved within different theoretical tools, their differences (in the treatment of noun stress) can be easily quantified on a corpus of English words. Second, it is not difficult to come up with words that do not follow the stress rules of HV87 and H98. For example, *kangaroo*, a core pattern under HV87, is impossible under H98, which does not assign primary stress to the final syllable at all. Traditionally, exceptional words are annotated with diacritics or are to be "lexicalized," but I am not aware of any proposal on what counts as exceptions — other than stipulating that they are exempted from the core stress rules. The Tolerance Principle promises a resolution for this longstanding problem: if these theories of stress are to regarded as an accurate — and productive — characterizations of the English grammar, then the number of exceptions must fall below the requisite threshold.

4.2.2 Prosodic Development and Learnability

Children must have acquired considerable knowledge of the phonology, morphology, and syntax of English before successfully acquiring the system of metrical stress. First, they need the segmental inventory of the native language, which is typically fairly complete before their first birthday (Werker and Tees 1984 as well as Kuhl et al. 1992; see Yang 2006a for a general review). Second, they need the basic phonotactic constraints of the language (Halle 1978) in order to construct syllables that are the building blocks of the metrical system. For instance, Dutch- and English-learning infants at nine months prefer consonant clusters native to their languages despite the segmental similarities between these two languages (Jusczyk et al. 1993b). Third, children must be capable of extracting words from continuous speech, certainly no later than at seven and half months (Jusczyk and Aslin 1995), which may exploit both statistical and structural properties of the lexicon (Bortfeld et al. 2005; Chomsky 1955; Harris 1951; Johnson and Jusczyk 2001; Jusczyk and Hohne 1997; Lignos 2013; Saffran, Aslin, and Newport 1996; Yang 2004). Finally, children must be able to detect prominence of stress, even though it remains an open problem how they may do so on the basis of acoustic information (Sluijter and van Heven 1996). Indeed, very young infants appear to have identified the statistically dominant stress pattern of the language, because 7.5-month-old English learning infants perform better at recognizing trochaic than iambic words (Jusczyk, Cutler, and Redanz 1993a; Jusczyk, Houston, and Newsome 1999). At a minimum, the child is able to locate primary stress on the metrical structure of words, and acquisition of the metrical system probably starts well before the onset of speech. I return to the issue of trochaic preference in early child language, because it appears to be a transient stage en route to the target grammar. Overall, these developmental preliminaries are supported by the current understanding of perception and speech development in children and, as will be clear, are indispensable for any treatment of stress acquisition.

I assume, along with other researchers (Dresher and Kaye 1990; Fikkert 1994; Hayes 1995; Idsardi 1992), that Universal Grammar provides a core set

4.2 A Recursive Approach to Stress

of parametric options that delimit a range of possible metrical structures (syllable, feet, etc.) and possible computational operations (e.g., projection, foot building, edge marking) that manipulate these structures. But it is inconceivable that the *totality* of crosslinguistic variation is innately available to learners. Rather, I envision learners experimenting and evaluating the core metrical hypotheses in an incremental fashion as they process and internalize the linguistic data.

(14) a. If a grammar fails to reach productivity, it is rejected, which triggers learners to revise their grammar.

b. If there are multiple grammars meeting the Tolerance threshold, learners select the grammar with the fewest exceptions.

c. If no grammar is productive, the stress patterns of words are memorized as a lexicalized list.

Each grammar G_i, a system of stress assignment, is to be associated with a tuple (N_i, e_i), the number of exceptions (N_i) it could apply to, and the number of words that contradict it (e_i). Thus, learners traverse through a sequence of grammars following a trajectory below.

(15) $G_1(N_1, e_1) \to G_2(N_2, e_2) \to G_3(N_3, e_3) \to \ldots \to G_T(N_T, e_T)$

Under this view, G_{i+1} is more highly valued than G_i, as the result of the additional data accumulated during the time between G_i and G_{i+1}. In particular, the additional data may have rendered G_i unproductive — more exceptions appeared — thereby forcing learners to adopt a different grammar G_{i+1}.[7] Here learners follow a developmental path quite similar to the emergence (and demise) of the inflectional rules studied in section 4.1. A grammar's productivity may change as a function of N_i and e_i. It is also possible that UG provides certain markedness hierarchies, which lead learners to entertain some grammars before others. For instance, it is conceivable that quantity-insensitive systems are simpler than quantity-sensitive systems, and learners consider the latter option only if the former has been rejected by the Tolerance Principle.

---

[7] It is possible for learners to "loops back" to a grammar rejected earlier. For instance, a child may draw a sample of ten words that have no exceptions, which results in a completely productive rule R. An additional ten words are learned, none of which follow R, and the learner would have to lexicalize all twenty items. Finally, an additional fifty words are learned, all of which follow R and now R is back in business because the the exceptions are well below the critical threshold of $\theta_{70} = 16$. Thus different learners may follow somewhat different paths of grammar traversal but as long as their input data is largely comparable, they can converge to the same terminal grammar.

To operationalize the conception of learning in (15) I first construct an approximate sample of the child's vocabulary and then evaluate the two leading theories of the English metrical system (HV87 and H98) reviewed earlier. Similar to the earlier studies considered in this chapter, I use a child-directed English corpus to approximate the input to learners.[8] I only evaluate the words that have been automatically tagged as common nouns and verbs. Since nouns and verbs have somewhat different stress patterns, considering them together will pose a realistic test for any model that seeks systematic regularity amid a heterogeneous mix of patterns.

In some of the studies below, words are morphologically processed using a computerized database from the English Lexicon Project (Balota et al. 2007). Morphology is well known to play an important role in English stress assignment (Chomsky and Halle 1968) and, as I show momentarily, must be taken into account for acquisition to succeed. The crosslinguistic study of morphological acquisition (reviewed in chapter 2) has made it very clear that inflectional morphology is acquired relatively early. This justifies our assumption that young children can take the inflectional structure of words into account during the acquisition of stress; for example, they may realize that *walks*, *walking*, and *walked* are all variants of the verb *walk*.

In all our studies, an electronic dictionary (the CMU Pronunciation Dictionary 2008) is used to obtain the phonemic transcriptions of words, which are then syllabified following the Maximize Onset principle (Kahn 1976), with sonorants and glides in the coda treated as syllabic.[9] A long vowel (diphthongs and the tense vowels) is regarded as heavy (VV); otherwise it is light. The HV87 and H98 systems are implemented as computer programs that take a word (a sequence of syllables) as input and assign primary stress to it. The output of the programs is then compared to dictionary stress assignment so as to evaluate the accuracy of the stress theories—specifically, tallying up the

---

8 The acquisition of stress presented here was based on an earlier project in collaboration with Julie Anne Legate, part of which appeared as Legate and Yang 2013. Here we used a slightly smaller, 4.5-million-word corpus of child-directed English from the CHILDES database; the current 5-million-word corpus included materials deposited in the years since.

9 Entries that could not be found in these lexical databases are omitted. These are almost exclusively transcription errors or nonsense words in the CHILDES database. The CMU pronunciation dictionary does not contain part-of-speech information, making it impossible to distinguish the homographic words with distinct stress patterns (e.g., *recórd* the verb and *récord* the noun). I used the CELEX database, which has part-of-speech annotation, in conjunction with the pronunciation dictionary to obtain the correct transcription.

number of exceptions. Since we are only concerned with the acquisition of primary stress, other levels of stress are ignored. For instance, the stress contour of a trisyllabic word such as *animal* (in its dictionary form) will be represented as "100" where 1 stands for primary stress and 0 otherwise.

4.2.3 Stage Transitions in Stress Acquisition

A thorough treatment of stress acquisition would measure the incremental growth of the learner's vocabulary and evaluate alternative grammars along the way. That is far too complex for the current undertaking. For simplicity, I only consider two representative stages of stress development, one designed to capture the child's stress system under a very small vocabulary and the other when the child has already learned enough words to potentially match the target state.

4.2.3.1 An Early and Quantity-insensitive Grammar

I first consider only words that appear more than once per 10,000 words, resulting in a set of 420. Most of these are, as one might expect, very simple words that young children experience in their daily life. The distribution of stress patterns is summarized in Table 4.3.[10]

Table 4.3
Stress patterns for words with frequency $\geq$ 1 in 10,000. Only 10 words (in boldface) do not stress the initial syllable and would be regarded as exceptions to the initial-stress rule.

Contour	Counts
1	287
10	107
100	13
01	7
010	3
1000	3

---

10 The three extraordinarily long words are the four-syllable *everybody*, *anybody*, and *caterpillar*.

The statistical distribution of stress assignment in Table 4.3 is clearly consistent with a quantity-insensitive system that stresses the first syllable: learners only need to deal with 10 exceptions, far below the tolerable threshold of $\theta_{402} = 67$. Interestingly, children learning English and similar languages do go through an initial stage during which the child is limited to a maximum bisyllabic template with the primary stress falling on the first. Fikkert 1994 is the most detailed longitudinal study of stress acquisition in the literature. Fikkert notes that children learning Dutch, a quantity-sensitive language, nevertheless start off with a quantity-insensitive grammar. They tend to stress the initial syllable in disyllabic words; for words in which the primary stress falls on the final syllable, children often shift it to the initial syllable as in *ballòn→bàllon* 'ball', *giràf→gìraf* 'giraffe', etc. Moreover, when the few trisyllabic words are reduced to a bisyllabic form, the stressed syllable is always preserved as in *vakàtie→kàntie* 'holiday' and *òlifant→òfant* 'elephant'. The statistical distribution of Dutch stress appears to support the numerical dominance of a trochaic system (Daelemans, Gillis, and Durieux 1994), and the advantage is likely much stronger in shorter and more frequent words that are part of a young child's vocabulary.

More directly relevant to the current topic, the privilege of the initial syllable is also observed in children learning English; for reviews see Demuth 1996; Gerken 1994; but also Vihman, Depaolis, and Davis 1998. In Kehoe's 1997 study, toddlers are instructed to repeat words, including novel words, after adult speech. Across all age groups (from 22 to 34 months), children show a general tendency to preserve the (primary) stressed syllable in their production. Of special interest is the contrast between multisyllabic words such as *dinosaur* and *kangaroo*. If the initial-stress grammar is productive, then *dinosaur*, which stresses the initial syllable, conforms to the rule. In comparison, *kangaroo*, which stresses the final syllable, would be an exception. In fact, children across the age groups made considerably more errors on the *kangaroo* type than the *dinosaur* type (Kehoe 1997, table 3), and across all age groups, children have a tendency to shift stress to the initial syllable (Kehoe and Stoel-Gammon 1997).

The preference for an initial-stress system is probably not surprising since it is well known that English-learning children's early language has a large number of nouns (Tardif, Shatz, and Naigles 1997), most of which are monosyllabic or bisyllabic, thus heavily favoring the trochee. Of course, English is *not* a quantity-insensitive language; structural factors such as lexical category and morphological structures also affect the assignment of stress. I now explore the development of the stress grammar when learners acquire a larger vocabulary.

4.2.3.2 The Bifurcation of English Stress

When we expand the vocabulary for learning, the initial-stress grammar breaks down, prompting learners to revise their hypothesis. To this end, I consider words that appear at least once per million in the corpus of child-directed English, again focusing only on nouns and verbs. There are 4,047 nouns, 2,402 verbs, and 5,763 lexically and prosodically distinct words altogether.[11] We have no precise measures of children's vocabulary but given previous estimates for school-age children (Anglin 1993, 62) and the projections of vocabulary growth from earlier stages (Fenson et al. 1994), a vocabulary of some 6,000 words seems reasonable for a three- or four-year-old English-learning child.

Now the initial-stress grammar is no longer productive, even though it still accounts for an overwhelming majority of words (4,960, or 86%, i.e., 803 exceptions): the critical threshold for exceptions is $\theta_{5763} = 666$. Thus, I predict that even for learners with fairly modest vocabularies, the simple initial stress grammar, one that a toddler appears to use, can no longer tolerate the exceptions. A reanalysis of word stress is called for.

I take this prediction of the Tolerance Principle to be a nontrivial result. Indeed, one wonders why English learners do *not* persist with the quantity-insensitive treatment of stress, eventually resulting in a metrical system that ignores the syllable structure in stress assignment. After all, such a system would work remarkably well from a statistical point of view: only 14% of exceptions need to be lexically committed to memory. The numerical advantage for initial stress has long been noted for English words; Cutler and Carter (1987), for instance, found that 73% of word types in an electronic dictionary receive the primary stress on the first syllable and the ratio climbs to 88% in spoken English, with fewer than 10% of polysyllabic words having a weak initial syllable.

---

11 Words like *control* and *record* that appear in the input as both nouns and verbs contribute to both the noun and the verb counts; these will be used when learners evaluate distinct grammars for nouns and verbs. In the case of *control*, the word only contributes once to the total count of words since the noun and verb forms of *control* are metrically identical. A word like *record*, however, counts twice in the total word counts, since the verb and noun forms of the word are distinct.

No linguist, and certainly no English speaker, appears content with statistical dominance, not even an overwhelming one.[12] While it is difficult to point on the exact age at which the reorganization of the stress system takes place, it is clear that English learners *do* acquire a quantity-sensitive system under which nouns and verbs receive different stress assignments, probably quite early. Experimental studies have consistently found the effect on the role of syntactic category and syllable structure in the stress assignment of novel words (Baker and Smith 1976; Baptista 1984; Kelly 1992; Ladefoged and Fromkin 1968; Trammell 1978). In one study, Guion et al. (2003) asked English speakers to "glue" together two aurally presented syllables such as *tar* and *minz*, *pou* and *tist*, *be* and *lin*, etc., which resulted in several different types of syllable structures. The conjoined word was placed in different syntactic contexts (e.g., "I'd like a tarminz" vs. "I'd like to tarminz"), which forced the word to be used as a noun or a verb. The stress assignment by the experimental subjects showed very clear sensitivity to both the syllable structure and the part of speech of these novel words: a quantity-insensitive grammar would predict uniform stress assignment across the board. A follow-up study with young children (Oh, Guion-Anderson, and Redford 2011) shows that five-year-olds already take parts of speech as well as syllable structure into account in stress assignment.

Thus, at some point in language learning, the children must have abandoned the initial-stress grammar, which has failed the Tolerance Principle. What grammar should be the replacement? In the limit, a child could decide to lexicalize everything (i.e., thousands of words). This is not a priori impossible; after all, every word we learn, and there are thousands of them, is an arbitrary mapping between sound and meaning. But at the same time, to the extent that a language can generate an infinite number of words, certain regularities for stress assignment must be discovered. This is where the Principle of Maximize Productivity, (13) in chapter 3 and repeated below, becomes relevant:

(16) *Maximize Productivity*
Pursue rules that maximize productivity.

---

[12] It remains possible that, for adults, the initial stress serves as a statistical cue for word boundary detection when no other cues are available; see Cutler 1996 for a review. The use of this strategy by infants is equivocal, however, because their treatment of stress undergoes developmental changes in the first year of life (Jusczyk, Cutler, and Redanz 1993a; Jusczyk, Houston, and Newsome 1999).

To obtain a different and productive grammar, learners have several moves to make. One possibility is to discover regularities within separate lexical classes (e.g., nouns and verbs). Children are well prepared to undertake this task, because the knowledge of syntactic categories is acquired extremely early and accurately (see, among others, Shi and Melançon 2010; Valian 1986; Yang 2013a). Although the entire vocabulary fails to support a productive grammar, fracturing the data into two distinct sets may identify productive grammars within: recall that the proportion of tolerable exceptions is higher for smaller numbers of N (Table 3.1). Another possibility is to consider the interaction between morphology and stress. English inflectional suffixes do not trigger stress shifts in the stems but some of the derivational affixes do (e.g., *-ic* but not *-ment*). The role of morphological learning in stress acquisition deserves additional discussion.

An English-learning child is well positioned to take inflectional morphology into consideration in the computation of stress. All inflectional suffixes are learned before 3;6 according to Brown's (1973) 90% obligatory usage criterion, and it is likely that these suffixes are reliably put into use in comprehension even earlier: children as young as twenty months understand the meanings of inflected verbs (Golinkoff et al. 1987) and by their second birthday, they can use morphological information to deduce the meanings of novel words (Naigles 1990). As argued in section 4.1, inflectional morphology is highly frequent in the input which facilitates early acquisition. Derivational affixes, however, are an altogether different matter. They are not fully acquired until much later, perhaps only during the school years (Jarmulowicz 2002; Tyler and Nagy 1989). For the purpose of stress acquisition, then, children are most certainly capable of "lemmatizing" inflected words to a single stem but are probably still unable to recognize derivational relations among words (e.g., *govern-government, real-reality*); for the acquisition of derivational morphology see section 4.3. In other words, it is reasonable to assume that young children are capable of treating all inflectional forms of *walk* (i.e., *walk, walks, walked* and *walking*) as *walk* for the purposes of stress acquisition. From the perspective of Tolerance, this has the effect of reducing the relevant vocabulary size (N), which enhances the likelihood of discovering a productive grammar. Furthermore, I assume that children have correctly learned that inflectional suffixes do not trigger stress shift, which is easily accomplished by the use of the Tolerance Principle because inflectional morphology never modifies the stress of the stem.

The process of learning can be visualized in Figure 4.2. Each step in the path of learning represents a more complex grammar, which is invoked only if the current, simpler, grammar fails to meet the Tolerance test. Presumably, a child learning a true quantity-insensitive language needn't go beyond the very first stage: the dominant stress pattern satisfies the Tolerance Principle over the entire vocabulary and no further subdivision is necessary. The next stage is to partition the lexical items into nouns and verbs. If productivity is satisfied, then the learner stops. Otherwise the learner looks to break up the data (within nouns and verbs) even further. For instance, eliminating the inflectional endings, which reduces the lexical set even further, will increase the chance of reaching productivity.

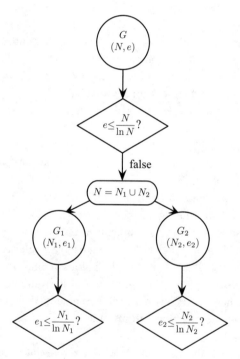

Figure 4.2
Learners may partition the lexical set upon failure to identify a productive rule. The Tolerance Principle will be used recursively within the subdivisions, which may lead to different stress rules.

I now subject to the theories of HV87 and H98 to a quantitative Tolerance test. Table 4.4 summarizes the results of evaluating HV87 and H98 under two

4.2 A Recursive Approach to Stress

conditions with respect to inflectional decomposition (stem±) and lexical separation (lex±). When evaluating grammars without making the lexical distinction ([lex+]) between nouns and verbs, I use the noun rules in HV87 and H98. Since the vocabulary consists of more nouns than verbs, the failure of the noun rules to reach productivity over the entire set of words means that the verb rules have no chance either. It is quite clear that a monolithic treatment of stress regardless of lexical categories cannot be correct, and that the English grammar must develop different rules for nouns and verbs. For a grammar with separate rules for nouns and verbs, it is only considered to be successful if its rules reach productivity for both nouns and verbs.

Table 4.4
Evaluation of stress grammars for words with frequency ≥ 1 per million. (a) with 515 exceptions. (b) with 355 exceptions.

lex	stem	HV87	H98
−	−	no	no
−	+	no	no
+	−	no	yes[a]
+	+	no	**yes**[b]

The H98 system under (lex+, stem+) is the best grammar under evaluation. H98 under (lex+, stem-) also manages to reach productivity, but it must lexicalize more exceptions and is thus less favored. It is interesting to examine the exceptions according to the H98 system. Many exceptions are nouns with a final syllable containing a long vowel, which ought to receive final stress according to H98 (see the description in (13)) but do not. On inspection, most of these end in the long vowel /i/, including the final diminutive suffix (*kitty*, *doggie*, *birdie*) as well as morphologically simplex words such as *body*, *army*, and *monkey*. Halle (1998) notes (see also Liberman and Prince 1977) that these suffixes are unstressable and are therefore ignored by the rules for stress assignment. Although he does not address how learners might reach such conclusions, the Tolerance Principle can be straightforwardly applied to detect regularities within exceptions. The morpheme segmentations in the English Lexicon Project lists 530 word with *-y* suffix: none receive primary or even secondary stress. Thus, the unstressability of certain segments is in principle learnable from the language data. If so, the empirical coverage of the H98 system will be further enhanced.

To conclude this study of English metrical stress, it is important to note that I have merely *evaluated* the stress rules hypothesized by linguists (Halle 1998; Halle and Vergnaud 1987; Hayes 1995). We know little about how children, constrained by the universals and parameters of the metrical system, *construct* these rules, including all the language-specific idiosyncrasies. Nevertheless, I hope to have made a convincing case that the Tolerance Principle can be an effective guide, steering children clear of excessive memorization and directing them toward the effective rules of grammar. Furthermore, I hope to have demonstrated the utility of quantitative evaluations of linguistic theories. In many cases, the empirical differences are often subtle and can only be magnified when a sizable body of data is taken into account.

A final note before moving on. The incremental acquisition of language is perhaps a necessary condition for the emergence of the English stress system. On a relatively small set of words — say, a few thousand — the structural generalizations of stress assignment receive adequate statistical support, because the number of exceptions can be kept beneath the tolerance threshold. But no rule is likely to survive the statistical assault of a very large vocabulary. For instance, if all 130,000 plus words are taken into account, some of which are very rare, even the statistical dominance of initial stress is considerably weakened (to only 71%, as opposed to 85% in the child-directed corpus). Even Halle's 1998 theory, the most accurate description of English stress to date, fails the Tolerance test. But does it mean that there is no regularity to English stress at all? Evidently not: How else do we account for the systematic patterns in stress assignment by speakers of English, children and adults alike? The only sensible conclusion is that the stress grammar is by and large fixed at a fairly young age when the vocabulary is small and the tolerance threshold relatively high (Table 3.1), and the late-arriving items do not become part of the effective vocabulary. These conjectures have interesting and potentially important implications for how language and language acquisition are situated in a broad context of human cognition and development, to which I return in chapter 7.

4.3 The Mysteries of Nominalization

In this section, I tackle one of the "central mysteries" in word formation (Aronoff 1976, 35): the derivational morphology of English. Traditionally, it

4.3 The Mysteries of Nominalization

has been convenient to regard inflectional and derivational morphology as fundamentally distinct processes, where the former is presumed to be more productive than the latter. But this is surely only a matter of tendency, and should not be elevated to a theoretical principle or regarded as a deep property of linguistic systems. Some aspects of derivational morphology, such as the nominalizer *-er* as in *dream-dreamer*, are extremely productive and consistently apply to new words that enter the language. At the same time, inflectional morphology can also be lexicalized and unproductive, with the English irregular verbs/rules an obvious case in point. Even more significantly, the existence of morphological gaps in the inflectional system (chapter 5) suggests that the absence of productivity is not an exclusive feature of the derivational domain, as Halle (1973) noted long ago.

4.3.1 Productivity and Frequency

For the present study, I focus on a selection of nominalization suffixes that have been extensively discussed in the past literature (Anderson 1992; Aronoff 1976; Baayen and Lieber 1991; Chomsky 1970; Embick and Marantz 2008; Halle 1973; Halle and Marantz 1993; Jackendoff 1975; Lieber 1980; Marantz 1997, 2001; Marchand 1969; Siegel 1974). I start with their descriptive characteristics on the basis of Plag's (2003) treatment; these are chosen because they represent both productive and unproductive processes in nominalization, as well as those that appear to fall somewhere in between.

(17) *-age*: derives nouns that express activity or result (*coverage, leakage*), and nouns that denote collective entity or quantity (*acreage, voltage*)

-er/or: derives nouns that signify active or volitional participants of events (*teacher, singer*), nouns that denote entities associated with instruments (*blender*) and activities (*trainer*), as well as persons or places of origin or residence (*Berliner*)

-ity: derives nouns that denote qualities, states, and properties (*productivity, solidity*), along with a fairly lexicalized set that have idiosyncratic meanings (*antiquity*); can productively attach to the suffixes *-able, -al*, and *-ic* and words ending in [ɪd] (*readability, formality, electricity*, and *solidity*)

-ment: derives action nouns that denote processes or results from (mainly) verbs with strong preference for monosyllable or disyllabic base words with final stress (*assessment, treatment*)

-ness: a suffix that can attach to most adjectives (*calmness, usefulness, impressiveness*)

-th: completely lexicalized; nominalization that may also trigger unpredictable stem change (*length, width, growth, warmth*)

There are two important aspects of this study I wish to highlight before proceeding. First, the relevance of the stress-acquisition study is obvious. If learners were to try any of these suffixes as a "global" default for nominalization, clearly none would stand a chance. According to my corpus analysis — details momentarily — the suffixes in (17) have the type frequencies of 6 (*-th*), 22 (*-ness*), 24 (*-age*), 27 (*-ity*), 45 (*-ment*), and 354 (*-er/or*). For a total of $N = 478$ derived nominals, an overall default suffix needs at least 401 members because the threshold for exceptions is $\theta_{478} = 77$. Another immediate observation is that (type) frequencies of these suffixes do not correlate with their productivity in any clear way — for instance, *-ness* is the second least frequent suffix but is generally regarded as the most productive among the entire set.[13]

Note that a default process for all nominalization is certainly a possibility among the world's language. In Mandarin, for instance, the nominalizer *de* is applicable across the board:

(18) a. fei de
fly DE (what flies)

b. hong de
red DE (what's red)

c. shangmian de
above DE (what's above)

d. zhangshang chi de
morning eat DE (what's for the morning, i.e., breakfast)

Clearly, this strategy will not even get off the ground for English. Thus, the Tolerance Principle will immediately force learners to fracture the nominalization process into subclasses and try to establish productive rules within. These

---

13 Nor do they correlate with the hapax legomena-based productivity metric proposed by Baayen and his colleagues (Baayen 1989, 2009; Baayen and Lieber 1991) over the years.

include the adjective-attaching suffix (*-ness*), the *ible*-adjective-attaching suffix (*-ity*), the verb-attaching suffix (*-ment*), etc., and the Tolerance Principle will be used recursively within each class, just like the partitioning of English words into nouns and verbs and seeking productive (but different) rules within each set. In the discussion that follows, I assume that learners have already rejected the possibility of a single nominalization default and are on course to discover rules within each suffixed subsets.

The second point concerns what may be called the ecological validity of language learning. Linguists are fond of dictionaries and, more common these days, large electronic corpora. These resources will continue to be invaluable for linguistic description and analysis, but they are of limited utility for the study of morphology acquisition. For instance, to the extent that the descriptions in (17) accurately reflect most English speakers' knowledge of derivational morphology, they should be "projectable" from a reasonable sample of English data where, again, distributional sparsity reigns supreme. Children shouldn't need to become professional wordsmiths to learn how to derive nominals in their native language. Hence I continue to use the five-million-word child-directed English corpus. It is my contention that the structural properties of the English nominal suffixes in (17) can be successfully acquired from everyday speech, although to do so obviously requires larger quantity of data (of comparable quality) than five million words. In chapter 7, I advance a further, and more radical, possibility that the grammar is *only* learnable with a relatively small sample of the language, much like the child-directed corpus or similar conversational corpora, and that learning in fact may *not* be possible if children start off with a college-level vocabulary or the OED.

4.3.2 Form Proposes, Meaning Disposes, Tolerance Decides

Let's outline the assumptions and mechanics for learning English derivational morphology. As discussed earlier, I assume that children have given up on a globally applicably suffix as in the nominalization system of Mandarin (18); the numbers are stacked against this possibility and the learners now seek to identify the specific properties associated with each suffix in (17).

I assume that the learners are suitably prepared, in a sense to be clear, on the phonological and syntactic/semantic ground to acquire morphology, which mediates the mapping between form and meaning in complex words. Suppose that the children encounter a selection of words (*fairness*, *silliness*, *teacher*, *reader*, *growth*, *warmth*, etc.) that are identified as nominals. First, I assume that the children are capable of morphological segmentation, in that they can identify phonological materials that are *potential* morphological units in the language. The term *potential* is important: how children learn to carry out morphological segmentation is not well understood, but there is clear evidence that they do. Laboratory studies have shown that even very young English-learning children can identify the morphological pieces such as those found in the complex Russian gender system (Gerken, Wilson, and Lewis 2005; see also Maratsos and Chalkley 1980). Furthermore, children actively break words up into pieces, occasionally resulting in segmentation errors. For example, Peters (1983) provides many instances of morphological misanalysis, including /mɛʒ/, a verb "to measure," evidently from the noun *measure* understood as an instrument for measuring as in "tape measure" (Weir 1962, 74). I will add to these with naturally occurring examples found in the CHILDES database in a moment. In practice, a distributional approach (Harris 1955) using transitional probabilities over phonemes does a passable job at least for languages such as English (Keshava and Pitler 2006), which is somewhat surprising because the method performs not nearly as well for word segmentation over syllables (Yang 2004).

A basic idea shared across all approaches to morphological learning is to identify morphological units or processes that relate a large number of paired words. A suffix is justified if it accounts for examples such as *teach-teacher*, *foolish-foolishness*, *govern-government*, and so on. Nevertheless, all computational models of morphological learning cannot escape certain glaring errors, even for state-of-the-art unsupervised learning systems (Lignos et al. 2009, 2010). For productive suffixes, there is a tendency to oversegment: *rubber* is often analyzed as *rub+-er* because *rub* is an actual word that can be found in the corpus. For suffixes with more restrictions, there is a tendency to underseg-ment; rarely are words such as *drainage* split into *drain* plus *-age*. My collaborative work on natural language processing suggests that productivity again holds the key to morphological learning. The model must be able to recognize which of the potential suffixes are actual suffixes, and which of the actual suffixes are productive and which are merely lexicalized. Subsequently, productive suffixes can be generalized to analyze novel data and lexicalized suffixes should not be extended beyond a finite and specific list. At the same time, syntactic and semantic consideration is also crucial so that the model is not misled into making spurious generalizations on the basis of form alone. An amusing example from a morphological learning model (Lignos et al. 2010) involves the postulation of a "suffix" *-et* that does, formally but spuriously,

4.3 The Mysteries of Nominalization

relate pairs of words such as *bull-bullet*, *dock-docket*, *wall-wallet*, and even *ass-asset*. Unless the model had some appropriate appreciation of meaning, it would be difficult to identify the nonrelationship between *wall* and *wallet* such that the pseudosuffix *-et* may be eliminated.

This is not the place to construct an automatic analysis of English morphology but algorithmic considerations in an unsupervised setting of language learning are nevertheless instructive. Suppose learners accumulate a reasonable number of paired words $\{w, w'\}$ whose forms are related by a potential suffix S (e.g., $w' = w \cdot S$). Now the status of S can be put to the test, with three possible outcomes:

(19)   a. S is a productive suffix (e.g., *fair-fairness*, *still-stillness*): the semantic relation between w and w' is systematic and extendable. No lexicalization is needed for w such as *fair* and *still* and S can be generalized to novel instances.

       b. S is an unproductive suffix (e.g., *warm-warmth*, *grow-growth*): the semantic relation between w and w' is systematic but not extendable. Lexicalize w with S such that *warmth* and *growth* are treated as morphologically complex, but S should be extended beyond the attested examples.

       c. S is spurious (e.g., *ass-asset*, *dock-docket*): the meanings of w' and w are unrelated. Words such as *asset* and *docket* are treated as morphologically simplex, and S is to be purged from the pool of potential suffixes.

Two points in (19) require elaboration. One is the notion of "relatedness." More specifically, I assume child learners know that *fair* and *fairness*, *warm* and *warmth*, *teach* and *teacher* are semantically related but *wall* and *wallet* are not. There is little doubt that children make these distinctions, being the excellent word learners they are. And it seems that such semantic considerations are required for learning morphology in any theoretical framework.[14] An interesting approach to pursue, though not one necessarily suitable for language acquisition, is to assess the semantic relatedness between words as a function of their distributional distance in context. This is a very traditional idea in linguistics (Chomsky 1955; Firth 1957; Harris 1954), recently popularized thanks

---

14 For example, see Marantz 2013 and Harley 2014 on how the semantics of w and w' must be taken into account in morphology under Distributed Morphology, arguably the most abstract theoretical framework in practice.

to the Big Data revolution and advances in computational linguistics (Baroni and Lenci 2010; Mikolov et al. 2013), although there are well-known limitations to its effectiveness in detecting certain linguistic relations (Redington et al. 1998; Rubenstein and Goodenough 1965). What those approaches excel at, it appears, is capturing the semantic relatedness between words, which is exactly what's needed to prune spurious relations identified by a distributional learner. Presumably, the semantic distance between *fair* and *fairness* would be shorter — because they *are* very closely related — than that between *wall* and *wallet*, which typically have nothing in common with each other and the "suffix" *-et* is purely accidental.

The other remark concerns, of course, productivity. Specifically, I assume that among the N pairs of words (w, w') formally related by S, a sufficiently large number of them must also be semantically related, in the sense defined above, to justify the productivity of S. More precisely, there can be no more than θ_N pairs that are semantically unrelated. The Tolerance Principle, then, is charged with the task of distinguishing the three types of morphological units in (19). For a suffix like *-ness*, there must be an overwhelming number of semantically related pairs among those related by form. For a suffix like *-th*, the number should fall below the threshold of exceptions. Finally, a "suffix" like *-et* is to be banished because none of the word pairs are related.[15]

In sum, I have outlined a morphological learning model where form proposes and meaning disposes. The potential morphological units that survive scrutiny become actual morphological units, with the Tolerance Principle as the gatekeeper.

4.3.3 Evaluating Nominalization Suffixes

As an illustrative example, let's first consider the relatively straightforward case of *-ness*. What happens if learners treat *-ness* as a productive suffix of English? They will have many positive examples (e.g., *fair-ness*) but will also occasionally go astray: *witness* isn't funny at all (*wit*) and *goodness* expresses exclamation rather than quality (*good*).[16] But overall, learners will observe

---

15 It should be noted that there is another variant *-ette* (evidently derived from French) that is probably a diminutive suffix (e.g., *kitchen-kitchenette*). These examples do not appear in the child-directed corpus but if they did, I assume that children can recognize the semantic relatedness of the words and treat *-ette* as an actual suffix which, incidentally, also receives stress.

16 To be sure, *wit* can be used as a verb in the rather idiomatic form of "to wit": *witness* was indeed formed from *wit* in Old English. But there is no chance for young learners to derive this

4.3 The Mysteries of Nominalization

that -*ness* works well enough to deserve its productive status, and words such as the noncompositional -*ess* words will be have to be lexicalized. Figure 4.3 illustrates this decision-making process.

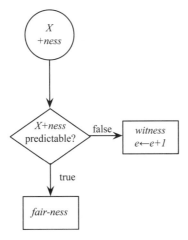

Figure 4.3
The tabulation of N and e to evaluate the productivity of the -*ness* suffix, where relatedness is semantically determined. Exceptions such as *witness* and *goodness* increases the count e.

The validity of a suffix is determined by the Tolerance Principle, and we again turn to our five million words of child-directed English. As shown in Figure 4.3, I extracted all words that contain a segmentable ending of -*ness*. If a -*ness* word is related to the word/stem to which -*ness* attaches (e.g., *fairness*), then it counts as a positive instance for -*ness* as a productive suffix. If, however, the presence of -*ness* does not contribute to the meaning in any clear way, as in the case of *witness*, it is chalked up as an exception. To operationalize the notion of relatedness, I largely relied on the morphological annotation provided in the English Lexicon Project (ELP; Balota et al. 2007), where almost all derivational nominals from our child-directed corpus can be found. The annotation in the ELP Corpus is fairly conservative and seems to follow the word-based tradition in morphology (Anderson 1992; Aronoff 1976) in emphasizing the role of full surface forms. For instance, words composed of

---

synchronically. In the 55-million-word SUBTLEX-US Corpus (Brysbaert and New 2009), there are 302 instances of *wit*—hardly a rare word—but all are nouns. Similarly, *goodness* may be used compositionally (e.g., *The logistic regression model has excellent goodness of fit*) but this usage is rare in everyday speech and is unattested in our child-directed corpus.

free stems (e.g., *treatment*) and bound stems (e.g., *sacrament*) are marked differently, and formally complex but semantically underived words (e.g., *witness*, *department*) are marked as morphologically simplex. I am not inclined to assess the competing morphological theories, but it does seem reasonable to suppose that, at least during the earlier stage of acquisition, children will largely rely on surface word forms to discover morphological rules. This is partly because the vast majority of the derivation nominals in our corpus *are* formally composed of surface word forms — except for the case of *-ity*, which I consider in detail below — and partly because the postulation of bound stems or even more abstract units requires justification from, ultimately, additional surface forms.[17] When a word cannot be found in the ELP annotation, which is rare, I consulted online dictionaries and other speakers to make decisions on the semantic relatedness between words.[18]

All words with the suffixal endings in (17) are extracted from the child-directed corpus and then manually inspected for correctness. For instance, in the study of *-ness*, I extract all words that have been identified as a noun by a part-of-speech tagger and end in /nɪs/. These will include words such as *fairness*, but also words like *witness* and *harness*, for they are nominals and also end in *-ness*. If the child is to test the productivity of *-ness*, *fairness*, which is morphologically segmented in the ELP Corpus, would count as a positive instance of *-ness* as a genuine suffix. By contrast, *witness*, which contains a stem/word *wit*, would constitute an exception because the combination of *wit* and *-ness* does not correspond to the meaning of *witness*. As such, *witness* is not morphologically segmented in the ELP. Words like *harness* (also *business*) lead to a different kind of misanalysis. Mechanically, removing *-ness* from these words results in a nonword/stem or even a bound stem (*har* and *busi* /bɪz/). I assume that, at least for the purpose of learning, these words are

---

17 For example, it is conceivable, and in fact has been proposed by morphologists, that *relig* is a root from which *religion* and *religious* are derived, and more radically, *ceive* is an abstract root whose meaning is contextually determined by the prefixlike material as in *con/per/re-ceive* (see Aronoff 1976, 102, who cites the (fictional) "great Semitic grammarian ben-Moshe (ms)," for an earlier discussion and Harley 2014 for a recent treatment in a completely different theoretical framework). But this type of analysis is motivated by the presence of words such as *religion*, *religious*, *conceive*, *perceptual*, and many others, such that the putatively shared abstract root may be regarded as an analytical possibility. It is very unlikely for a young child to entertain this hypothesis, not least due to the sparsity of morphological data reviewed in section 2.1.

18 I am grateful to Sarah Murphy for her assistance in this study.

4.3 The Mysteries of Nominalization

ignored for the calibration of -*ness* because the decision about compositionality does not arise when there are no pieces to combine.[19]

The numerical results for -*ness* are as follows. There are twenty-two -*ness* attached words in the child-directed corpus; these are exhaustively listed below:

(20) a. Related: coolness, cuteness, exclusiveness, fairness, fitness, foolishness, gentleness, greatness, happiness, madness, meanness, numbness, politeness, quietness, rudeness, sadness, sharpness, sickness, sweetness, thickness (20)

b. Not related: goodness, (eye)witness (2)

The numbers work in -*ness*'s favor. With N valued at 22, it can tolerate $\theta_{22} = 7$ exceptions but only 2 non-compositional, semantically unrelated, examples are observed. And all 20 in 20a are derived from adjectives. Learners can then confidently conclude that -*ness* is a genuinely productive suffix that, when used to derive new words, results in a predictable meaning.

The process of learning nominal morphology in Figure 4.3, and the example with -*ness*, follow essentially the same mechanics as the evaluation of grammars for inflectional morphology and metrical stress reported in the earlier sections. Learners detect structural relations in the words they encounter, which are then subject to the numerical test of productivity on the basis of semantic relations. With this method in mind, I turn to the other nominalization suffixes in English.

---

[19] It is possible that upon hearing a novel word such as *kruckness*, mature speakers, for whom the productivity of -*ness* is established, would be primed to identify *kruck* as an adjective. This is exactly like young children interpreting *sib* as an action on hearing *I am sibbing* (Brown 1957).

Consider now -*ment*, with the following words attested in the child-directed corpus:

(21) a. Related:
accomplishment, achievement, adjustment, advertisement, agreement, amendment, amusement, announcement, appointment, argument, arrangement, assignment, attachment, basement, compartment, development, disagreement, disappointment, encouragement, enjoyment, entertainment, entrapment, equipment, excitement, improvement, installment, investment, management, measurement, movement, pavement, payment, punishment, refreshment, reinforcement, replacement, requirement, statement, treatment, unemployment (40)

b. Not related: apartment, cement, comment, department, supplement (5)

The 40 nominals in (21a) are transparently derived from verbs. The 5 exceptions in (21b) require a brief commentary. Some (e.g., *apartment*) do not contain a verbal stem. Some (e.g., *department*) do contain a stem (free or bound), but one that does not contribute to the composite meaning. Still others contain a pseudostem that involves accidental homophony (e.g., *ce-/see* in *cement*, *com-/come* in *comment*) but also fails to contribute to the meaning of the complex word. In any case, none of the 5 words in (21b) can be said to be semantically related to the formally available "stems" after the segmentation of -*ment*. But these negative examples fall far below the tolerance threshold ($\theta_{45} = 11$). Thus, the productivity of -*ment* as a nominal suffix that attaches to verbs is learnable from the English corpus. Furthermore, the 40 verbs in (21a) are typically multisyllabic (35 out of 40), and the 5 exceptions involving *move*, *pave*, *pay*, *state*, and *treat* entered the language at least 500 years ago according to the OED. Furthermore, most of the -*ment* words were introduced after the seventeenth century with multisyllabic verbs. It is conceivable, then, that multisyllabicity was one of the conditions for the productive extension of -*ment*, which is reflected in the data here as well as in Plag's description of the suffix summarized in (17).

Moving along to -*age*, this suffix turns out to be unproductive. The following words have the form of a stem followed by -*age*:

4.3 The Mysteries of Nominalization

(22) a. Related: baggage, bandage, carriage, drainage, footage, leverage, luggage, marriage, mileage, package, passage, postage, shortage, storage (14)

b. Not related: cabbage, cottage, damage, forage, garbage, message, rampage, sausage, village, vintage (10)

All but one of *-age* attached words in (22b) contain a stem which either does not at all or only very opaquely contributes to the composite. For instance, clothing (*garb*) has nothing to do with trash (*garbage*) and *saus/ce* does not usually go into a *sausage*. The semantic relatedness between the stem and the words in (22a), by contrast, is very transparent. If learners are to treat *-age* as a productive suffix, then, they would encounter just as much noise (22b) as signal (22a), which naturally fails the Tolerance test ($\theta_{24} = 7 < 10$). Granted, the judgment in (22) is subtle but the abundance of the negative examples in (22b) makes it very likely that learners will reject *-age* as a productive suffix. This is a desirable outcome. As Chomsky notes in his "Remarks" (1970), nonproductive suffixes such as *-age* can — though of course do not have to — introduce idiosyncratic meanings, as the contrast between *marriage* and the fully productive gerundive form *marrying* shows.

The evaluation of *-th* as a potentially productive suffix must be a source of amusement for learners (as was for the present author). The entire set of *-th* suffixed nominals in Modern English are listed below; those in boldface appear in our child-directed corpus:

(23) broad+th → breadth
deep+th → depth
grow+th → **growth**
heal+th → **health**
long+th → **length**
steal+th → stealth
strong+th → **strength**
true+th → **truth**
warm+th → **warmth**
wide+th → width

The stems here are phonologically, syntactically, and semantically diverse. They can be verbs (*grow*) or adjectives (*warm*), they can end in a vowel or consonant, and some undergo stem change while others don't — which I grant, for the sake of argument, does not prevent learners from recognizing the relatedness between, say, *strong* and *strength*. Presumably, this requires the same ability that recognizes the relatedness between words such as *catch* and *caught* in the past tense, where the modification to the phonology of the stem is equally arbitrary. But *-th* clearly cannot be a productive suffix. In the same child-directed corpus that contains the six positive examples of *growth*, *health*, *length*, *strength*, *truth* and *warmth*, learners are also confronted with *tooth* (not related to *too*), *teeth* (not related to *tee/a*), *booth* (not related to *boo*), *youth* (not related to *you*), *filth* (not related to *fil(l)*), *wealth* (not related to *well*), etc. The noise again overwhelms the signal, and *-th* must be lexicalized for the list in (23) and nothing more.

The behavior of *-er* (and its orthographic variant *-or*) and *-ity* is more complicated. Consider first *-er*, the suffix that generally denotes active or volitional participation but can also signify entities associated with instrumentation. The *-er* suffix is the only derivational morpheme tested in Berko's (1958) classic study. Despite the task-specific difficulty of the Wug test (see section 2.3), at least some of the children provided *zibber* for "the man who zibs." Therefore, the *-er* suffix must have become productive by a relatively young age. Clark and Hecht (1982) find that four-year-old children are able to segment *-er* to "underive" the stem verb; see also Clark and Cohen 1984, which shows that children acquire the *-er* suffix before school age. Moreover, productive use of *-er* can be found in the naturally occurring speech of young children, similar to patterns of past-tense overregularization. Some of the examples I found in the CHILDES database are given below, where the *-er* suffixed words are clearly children's creation:

(24) a. The *flatter* (referring to the rolling pin; 2;7).

b. She and Jenny took the *sounder* off with the needle (referring to an LP record; 4;6).

c. But you really call it the Darth Vader collection *caser* (referring to a container; 4;2).

d. It always sweats me. That *sweater* is a hot sweaty sweater (referring to the causer of sweating; 4;3, from Bowerman 1982).

And *pee-er* and *poo-er* are perennial favorites.

In any case, the productivity of *-er* is not in doubt for mature speakers. On the one hand, it readily attaches to new vocabulary items such as *blogger* (someone who blogs) and *redditor* (someone who uses Reddit). On the other hand, there is real-time processing evidence that *-er* is segmented off by

4.3 The Mysteries of Nominalization

English speakers even when doing so leads down a garden path. For instance, very brief visual presentation of *brother* induces a priming effect for *broth*, a pseudo-morphologically related stem (Rastle, David, and New 2004), but no effect is found for the orthographically equally similar *brothel*. The contrast is accounted for by the fact that *-er* is a productive suffix in English and speakers cannot help but parse it off in lexical processing (Anderson 1988b; Anshen and Aronoff 1988; Taft and Forster 1975); *-el* is not an actual morphological unit and triggers no comparable effect.

I again approach the status of *-er* from the Tolerance perspective: children need to tally up the number of derived words whose meanings are predictably derived with *-er* (e.g., *teacher*) along with those that do not (e.g., *rubber*). One must hope that eventually, the *teacher* type thoroughly outnumbers the *rubber* type. Therefore, I extracted all nouns in the child-directed corpus that end in a schwa or syllabic /r/ and then inspected them manually to determine the number of *teacher* and *rubber* types. There are 314 transparently compositional *-er* words, too numerous to list here exhaustively, that are clearly related to the stem with the semantics of agency or instrumentation. But there are also 40 noncompositional forms, all of which are annotated as simplex in the ELP corpus or contain a pseudostem. These are exhaustively listed below:[20]

(25) banner, bugger, bumper, buster, butter, cellar, chatter, cider, collar, copper, corner, counter, error, flicker, gutter, hammer, hangar, letter, lever, liquor, liver, manner, matter, meter, mister, professor, pucker, razor, rover, shoulder, sorcerer, summer, tailor, taper, tender, tractor, turner, whimper, whisker, whisper

Again, the judgment of the noncompositional examples is subtle. But for $N = 314 + 40 = 354$, the maximum number of exceptions is $\theta_{354} = 60$, which far exceeds the list in (25). In other words, treating *-er* as a genuine suffix does occasionally lead learners astray: *cellar* is not someone who *sells*, *banner* is not about prohibition, *counter* (usually) does not involving counting, etc. After failing to relate these pairs of words, learners will need to treat them as morphologically simplex, like *witness* in the analysis of *-ness* in Figure 4.3. Overall, however, the vast majority of *-er* nominals have meanings transparently derived from the stem; the *-er* suffix passes the Tolerance test.

---

20 I included *collar*, which could be misinterpreted as someone who *calls*, for speakers of the dialect that merges the vowels /ɑ/ and /ɔ/.

Finally, I take on *-ity*, a suffix that has featured considerably in the theoretical discussion of productivity (e.g., Aronoff 1976; Marantz 2001). Its complexity is manifest in both form and meaning (17). I begin by first listing the 27 *-ity* attached words in the child-directed corpus, again dividing them into semantically related (26a) and unrelated (26b) sets:

(26) a. Related: ability, activity, cavity, curiosity, electricity, humidity, infinity, insanity, morality, obscenity, possibility, priority, probability, purity, reality, responsibility, security, stability, stupidity (19)

b. Not related: amenity, gravity, nativity, personality, posterity, university, vanity, virtuosity (8)

Matters here are complicated. Again, the judgment is subtle. Moreover, as a nonneutral suffix, *-ity* frequently results in vowel change and stress shift on the stem (e.g., *curious-curiosity*), which may cause learners additional difficulty in morphological segmentation. This would partially account for the lateness in its acquisition, probably around the age of ten (Jarmulowicz 2002, 199, Tyler and Nagy 1989, 655).

A direct application of the Tolerance Principle to the *-ity* data in (26) suggests that its productivity is just on the cusp of productivity: the 8 negative examples just clear the threshold ($\theta_{27} = 8$). This is the desirable outcome, but it leaves the productivity of *-ity* in a pretty precarious state. The suffix, of course, is uncontroversially productive (Anshen and Aronoff 1988), as can be seen in novel words such as *friendable-friendability* in reference to the social network. A finer-grained analysis, and a recursive use of the Tolerance Principle, reveals a much stronger pattern. Many of the stems in (26) are adjectives, or are almost always used as adjectives, upon the segmentation of *-ity*, as shown in (27). The examples where the adjective stem does not transparently determine the semantics of the *-ity* composite are marked in boldface:

(27) able, active, curious, electric, humid, infinite, insane, moral, **native**, obscene, **personal**, possible, prior, probable, pure, real, responsible, secure, stable, stupid, **virtuosity**

Of the 21 adjectival stems here, 18 transparently contribute to the semantics of the derived words, all those that end in *-ible*, *-ic*, and *-al*. The three negative examples are well below the tolerance threshold ($\theta_{21} = 6$). If learners pursue this route, they can easily acquire the property of *-ity* as a suffix that productively attaches to adjectives of a certain phonological shape; see the descriptive summary in (17). The opaque use of -*ity* in (26b) can be relegated to lexicalization.

The current study, I hope, has shed some light on the mystery of English derivational morphology. Starting with the bare essentials of phonological and semantic considerations, the Tolerance Principle provides an independently motivated approach to productivity, which has been a focal point of controversy in morphological theory. I take it as an interesting fact that the structural properties of the nominalization suffixes can be projected from a relatively simple sample of spoken English and on an essential vocabulary that can be taken for granted for most if not all speakers of English. This calls for a very robust, and ultimately very simple, model of morphology and morphological learning, a theme I will continue to explore in the remaining pages.

4.4 The Horrors of German: Exceptions that Force the Rules

A major challenge for the dual-route model of morphology (Marcus et al. 1992; Pinker 1999; Pinker and Prince 1988) is that, despite the insistence on a rule-vs.-memorization dichotomy, it has never provided a successful strategy for identifying what makes a productive rule. Of course, *theorists* may know how to find the rule: Wug tests can be administered to native speakers, and the absence of word frequency effects in real-time processing is another hallmark of productive rules (e.g., Clahsen 1999). But these methods are verificational in nature and inapplicable to language acquisition; children do not query their parents for grammaticality, nor can they intuit lexical decision latencies. Unsupervised means for discovering rules are clearly needed. English inflectional morphology, the primary battleground in the past-tense debate, is straightforward: the rule/pattern that covers the most variety, or types, of verbs can be identified as the default, which correctly picks out -*d*. And it is in this context that the German plural system becomes prominent.

4.4.1 More Regularity After All

German marks noun plural with five suffixes: -*(e)n*, -*s*, -*e*, -*er*, and -*ø*, with the latter three also allowing umlaut (+UML) for some nouns. The basic patterns are summarized in Table 4.5.

Despite the very low frequency of the -*s* suffix, it is overused by German children (Clahsen et al. 1992; Marcus et al. 1995; Park 1978). It is also automatically attached to certain novel nouns that recently entered the German

Table 4.5

The German noun plural suffixes (Clahsen et al. 1992) and their frequencies from CELEX (Sonnenstuhl and Huth 2002)

Suffix	Singular	Plural	Gloss	Type	Token
-ø	der Daumen	die Daumen	"thumbs"	4320	87088
	die Mutter	die Mütter	"mothers"	(17%)	(29%)
-e	der Hund	die Hunde	"dogs"	6836	62239
	die Kuh	die Kühe	"cows"	27%	21%
-er	der Wald	die Wälder	"forests"	1067	10158
	das Huhn	die Hühner	"hens"	(4%)	(3%)
-(e)n	die Strasse	die Strassen	"streets"	12365	134492
	die Frau	die Frauen	"women"	48%	45%
-s	das Auto	die Autos	"cars"	1061	5468
	der Park	die Parks	"parks"	(4%)	(2%)

language (e.g., *die iPhones*). Thus a minority rule, in fact the smallest rule as shown in Table 4.5, can be productive, which presents a nontrivial problem for any account of productivity.

In a critique of Pinker 1999, and alluding to Mark Twain's (1880) well-known lament about the German language, I made the following assertion in *London Review of Books* (Yang 2000, 33):

> Pinker argues convincingly that, despite its low frequency, the *-s* is the default suffix. However, it's hard to believe that German speakers memorise [sic] all four classes of irregular plural, i.e. the majority of nouns in the language, on a word-by-word basis, as if each were entirely different from the others. The partial similarity among English irregular verbs, which looks like nothing more than a historical accident, has misled Pinker into looking for family resemblances: a quick glance at German shows that the four irregular classes of plural show no systematic similarity whatever. The horrors of German are real: one must sort each irregular noun into its proper class, as in the traditional rule-based view.

Chalk it up to youthful indiscretions.

Minority productive rules are simply impossible under the Tolerance Principle: productivity only obtains when the rule-following items thoroughly overwhelm the exceptions (Table 3.1). Thus, on numerical grounds alone, we are pressured into the position that at least some of the four "irregular" suffixes must be "regular" (i.e., productive). They must apply to some subsets of nouns, which in turn do not count as exceptions against the productivity of *-s*. This

4.4 The Horrors of German: Exceptions that Force the Rules   123

recalls the hypothetical examples considered in section 3.5.4 and repeated below:

(28) R_1: If $[+A, +B]$ THEN X
       R_2: If $[+A]$ THEN Y
       R_3: Z

It is possible for all three rules to be productive, namely, words with the feature $[+A, +B]$ automatically follow R_1, words with the feature $[+A]$ automatically follow R_2, and words that are otherwise unspecified follow R_3. R_3 would be considered the default because it is the least marked but it is clear that being productive is a necessary, but not sufficient, condition for a rule to become the default: R_1 and R_2 are productive as well but not defaults. The scopes of these rules, from more specific to more general, are implemented by the Elsewhere Condition, which ensures that R_1 is activated prior to R_2, which goes before R_3. In section 3.5.4, I also discussed the nontransparent relationship between the productivity of a rule and the type frequency of the words that it applies to. In (28), for instance, the rules may be applicable to 70, 20, and 10 words respectively, but they all are productive—that is, automatically applicable to words that meet their structural descriptions.

In this section, I show that the minority suffix -s in German is similar to R_3, a productive and least specified rule for plural formation. In contrast to the dual-route model, I demonstrate that the other suffixes, while subject to additional structural restrictions, are also productive, similar to R_1 and R_2 in (28). I approach the problem as someone not knowing the German language at all, which can be viewed as an approximation of how a German-learning child acquires the noun plural morphology. My analysis is mechanically forced by numerical calculations under the Tolerance Principle: the search for productive rules leads me to consider the role of phonology and grammatical gender in the formation of plurals, and the resulting account is very similar to the theoretical treatment of German morphology proposed by many other scholars (e.g., Bierwisch 1967; Dressler 1999; Mugdan 1977; Szagun 2001; Wiese 1996; Wunderlich 1999).

Before my quantitative analysis, let's peek ahead and briefly preview the evidence for the productivity of the non-s suffixes. Pedagogical and descriptive accounts of German morphology consistently portray the plurals as predictable, even though they are quick to point out the many exceptions. Statistical statements about the plurals make a similar point, because the following observation is typical: "For approximately 85 per cent of the nouns, masculine and neuter nouns take the plural -*e* or -ø, masculine nouns ending in -*e* and feminine nouns take -(*e*)*n*" (Elsen 2002, 117). That gender plays a role in plural formation can be seen in the following examples (from Wiese 1996, 139):

(29) a. das Partikel → Partikel, die Partikel → Partikel+n
b. das Steuer → Steuer, die Steuer → Steuer+n
c. das Koppel → Koppel, die Koppel → Koppel+n
d. der Kiefer → Kiefer, die Kiefer → Kiefer+n
e. der Leiter → Leiter, die Leiter → Leiter+n
f. das Mark → Mark+e, die Mark → Mark+en
g. der Flur → Flur+e, die Flur → Flur+en
h. der Marsch → Märsch+e, die Marsch → Marsch+en

The nouns in (29) have multiple genders: when it is non-feminine (left column), we find a plethora of suffixes, but when it is feminine (right column), as indicated by the article *die*, the -(*e*)*n* suffix is consistently used. Along a similar line, the productivity of the "irregular" suffixes can also be seen in patterns of variation and change. For instance, Elsen (2002, 117) notes that nouns (e.g., loanwords) that initially took the -*s* suffixes have been known to shift to "one of the other productive plural endings."

(30) Pizza/Pizzas → Pizzen 'pizza'
Kiosk/Kiosks → Kioske 'kiosk'
Modem/Modems → Modeme 'modem'
Balkon/Balkons → Balkone 'balcony'

This may happen to *die iPhones* someday.

More specifically, at least three major generalizations about plural formation can be identified (Wiese 1999; see Dressler 1999 for very similar observations):

(1) Feminine nouns predominantly take an -*n* as plural affix, whereas the plural form of nonfeminine nouns cannot be clearly predicted by gender alone. (2) Within the group of non-feminines, plural forms represented by the -*e* suffix are found as well as plurals marked with the -*er*, but the latter plural is in a clear minority. This view is confirmed by the countings based on the CELEX lexical database. (3) There is a substantial number of non-feminine nouns taking a zero plural. All of these are nouns ending in a so-called reduced syllable, as in *Filter*,

'filter,' *Segel*, 'sail,' *Garten*, 'garden.' Not a single noun consisting of just a single syllable or of two full syllables (see *Hund*, 'dog,' or *Arbeit*, 'work') ever has a zero plural. (Wiese 1999, 1044)

These statistical and distributional observations suggest that -*s* should not be singled out as the only productive suffix. In a moment, I will also review psycholinguistic and developmental evidence for the productivity of the non-*s* suffixes. For now, let's consider how a child may discover the plural rules in German.

4.4.2 How to Find Subregularities

As always, let's work through the input data that German-learning children receive and see how the regularities within the plural system can be identified. I collected a one-million-word child-directed German corpus from the CHILDES database (MacWhinney 2000) and extracted the 500 most frequent plural nouns, along with the gender and phonology of the stems and their corresponding suffixes. Two native speakers of German then went through this list and eliminated annotation errors as well as nouns that do not readily lend to pluralization; these include mass nouns such as *Butter* 'butter', *Honig* 'honey', and *Milch* 'milk' as well as several nouns that are almost always in the singular (e.g., *Bier* 'beer').[21] This results in a total of 458 nouns. Homophonous words such as *Leiter*, which is masculine when meaning "leader" but feminine when meaning "ladder" and takes different suffixes, and *Schloss*, which is neuter but can mean both "lock" and "castle," are counted separately as unique stems. While this is not a large corpus in terms of noun types, there are reasons to believe that a vocabulary of this size is what enables the acquisition of the plural suffixes. As we will see shortly, German children as young as 1;6 start to overregularize plural suffixes which means that they must have acquired their productive status by then. But an eighteen-month-old toddler cannot possibly have a very large vocabulary. While I am not aware of any systematic study of German children's vocabulary growth, research on the acquisition of English

---

21 I first ran a part-of-speech tagger (Schmid 1995) to extract the nouns. Then, using a pedagogical list for German noun-plural formation, the plural forms were extracted. This last step was necessary because no part-of-speech tagger, at least those in the public domain, distinguishes between singular and plural nouns. Beatrice Santorini then manually examined the list of nouns and verified their accuracy (for phonology, gender, and plural suffix). The words and the resulting analysis presented here have been further vetted by Florian Schwarz. I am grateful for their contributions to this study.

Table 4.6
Distribution of noun-plural suffixes for highly frequent nouns in child-directed German

Suffix	Types	Percentage
-ø	87	18.9
-e	156	34.1
-er	30	6.5
-(e)n	172	37.5
-s	13	2.8

suggests that the nouns in toddlers' language production are unlikely to exceed a few hundred (Fenson et al. 1994). Taking these considerations into account, the current, and relatively small, corpus is very suitable for the study of German noun-plural acquisition.

For the most part, I will only explore the choices of suffixation and put aside the issue of Umlaut, an independent process (lexically) triggered by suffixation (Wiese 1996), except in the case of monosyllabic neuter nouns that contain the vowel /a, o, u/. These nouns appear to productively take the *-er* suffix, which is always followed by Umlaut; see below. For comparison, I will also occasionally examine the distribution of noun plurals in a 900,000-word German newspaper corpus (TIGER; Brants et al. 2004), which has been manually annotated. Because the child-directed speech is relatively simple, virtually all noun plurals in our child-directed German corpus are monomorphemic (e.g., not compounds). The TIGER Corpus, on the other hand, does not provide compound marking so the word counts would include duplicates and should be used for reference purposes only.

The distribution of the suffixes in the child-directed German corpus is summarized in Table 4.6. For the child-directed data, we see that the relative proportions of the suffixes are quite similar to those reported from other sources (see Table 4.5 as well as Elsen 2002; Janda 1990; Szagun 2001). The statistics in Table 4.6 ensure that no suffix could emerge as the "monolithic" default. For a set of $N = 458$ nouns, a productive suffix can tolerate no more than 74 (θ_{458}) exceptions: an across-the-board suffix would need to account for at least 384 nouns and clearly none comes even close.

Again, I propose that the child follow the Principle of Maximize Productivity (16) to identify valid rules and generalizations by dividing the nouns into subclasses. Here gender is an obvious choice. German makes extensive

4.4 The Horrors of German: Exceptions that Force the Rules

use of gender throughout the language, and past research has found that children acquire gender marking very early. For instance, Mills's (1986) classic study shows that German children across all age groups rarely produce gender-marking errors. This is similar to the other cases of acquisition reviewed in chapter 2: most gender errors by children are those of omission rather than substitution. A more detailed study by Szagun (2004, 15) finds that the rate of correct gender marking is approximately 80% even before the age three, and it rises to nearly 100% before the age five. Interestingly, the 1986 study by Mills also notes that the acquisition of gender marking precedes, rather than follows, that of plural formation. This suggests that partitioning nouns according to gender is a logical as well as a developmental prerequisite for learning plural formation.

Therefore, the data summarized in Table 4.6 needs to be repartitioned, not by the suffix, but the gender of the noun. The suffixes for each resulting class will then be subjected to the Tolerance Principle for productivity evaluation. There are 166 feminine (*die*) nouns, 200 masculine (*der*) nouns, and 92 neuter (*das*) nouns; let's see the tabulation of productivity within each set.

4.4.2.1 The Feminine Nouns

Consider first the 166 feminine ([+fem]) nouns. Their distribution with respect to suffixation are described below.

(31) a. 146 take the -*en* suffix.
 b. 13 take the -*e* suffix: Angst 'fear', Bank 'bench', Ding 'thing', Fensterbank 'window sill', Frucht 'fruit', Gans 'goose', Hand 'hand', Haut 'skin', Lust 'pleasure, joy', Nacht 'night', Not 'need', Wand 'wall', Wurst 'sausage'.
 c. 6 take the -ø suffix: Ahnung 'guess', Gelegenheit 'opportunity', Metzgerei 'butcher', Pfadfinder 'boy scout', Stäbchen 'chopsticks', Weisheit 'wisdom'.
 d. 1 takes the -*s* suffix: Pizza 'pizza'.

For the [+fem] nouns, the choice between -*en* and -*n* is completely determined by the phonology of the noun: if the noun ends in a schwa, then -*n* is used; otherwise -*en* is used. This phonological alternation also holds for nonfeminine nouns as we shall see shortly. The statistics in (31) show that the -(*e*)*n* suffix has only 20 exceptions, whereas the tolerance threshold is at 32 (θ_{166}). I thus predict that -*(e)n* is the default suffix for [+fem] nouns; in fact, I am not aware of any treatment of German plurals that does *not* recognize this regularity, save the dual-route approach. The statistics from the TIGER Corpus further strengthen the conclusion of productivity. Of the 1,549 feminine plurals, 1446 take the -*(e)n* suffix, with 103 exceptions comfortably below the threshold of $\theta_{1549} = 210$.

The productivity of -*(e)n* for [+fem] nouns can be observed in several independent lines of research, in conjunction with traditional linguistic analyses (see the references cited earlier). First, when -*(e)n* attached [+fem] nouns are matched in stem frequency, lexical decision studies fail to find whole-word, or surface, frequency effects in reaction time (Penke and Krause 2002). The absence of whole-word frequency effects is uncontroversially taken as a hallmark for productive word-formation processes such as the English -*d* in the past tense (Pinker 1999), the German -*t* suffix in past participles (Clahsen, Eisenbeiss, and Sonnenstuhl 1997), and the German -*s* suffix in noun plurals (Sonnenstuhl and Huth 2002). Second, in the acquisition of noun-plural morphology, many studies have shown that children overgeneralize the -*(e)n* suffix at least as frequently as the -*s* suffix; if latter is regarded as productive, so should the former. Clahsen et al. (1992) report that a child named Simone overuses -*en* and -*s* equally frequently. Most other studies, however, report a stronger effect for -*(e)n*. Köpcke (1998), in a reanalysis of Clahsen's data along with some additional transcripts, finds nineteen instances in which -*en* was overgeneralized compared to only two instances of -*s* overgeneralization; similar results are obtained by Bittner (2000b), Park (1978), and Szagun (2001). In longitudinal studies, both Vollmann et al. (1997) and Elsen (2002) find that the -*(e)n* suffix — and the -*e* suffix, which I discuss momentarily — emerged earlier than -*s* during acquisition. Overgeneralizations can be observed before the age of two and are present throughout the course of language acquisition. In the corpus study of Elsen (2002, 121), -*(e)n* accounts for the majority of the errors (65% or 93/143), and even the -*e* suffix is overused at least as frequently as the -*s* suffix (25/143 and 23/143 respectively). More directly relating to the role of gender, Gawlitzek-Maiwald (1994) finds that the overapplication of the -*(e)n* suffix for feminine nouns is the dominant error throughout morphological development. All in all, we can conclude that the -*(e)n* is the productive plural suffix for [+fem] nouns, a fact that can be predicted numerically from child-directed linguistic data. The [+fem] nouns that do not take -*(e)n* need to

4.4 The Horrors of German: Exceptions that Force the Rules 129

be lexically marked for other suffixes, which are (a) subject to overregularization to -(e)n, and (b) predicted to, and in fact do, show whole-word frequency effects in lexical processing (Bartke et al. 2005; Penke and Krause 2002).

We now turn to the 200 masculine and 92 neuter nouns. Learners first consider the simpler analysis that treats these nouns in opposition to the feminine as [-fem]. Further partitioning them into [+masc] and [+neut] is pursued only when the [-fem] set fails to yield a productive rule, which turns out to be what's needed.

The distribution of suffixation for [-fem] nouns is more complicated than the [+fem] class.

(32) a. 143 take -e: e.g., Anruf 'call' and Geschenk 'gift'.
b. 81 take -ø: e.g., Esel 'donkey' and Zimmer 'room'.
c. 30 take -er: e.g., Bild 'photo', Blatt 'leaf', Gott 'god', Rand 'edge'.
d. 26 take -en: e.g., Affe 'monkey', Auge 'eye', Fleck 'spot', Hemd 'shirt'.
e. 12 take -s: e.g, Bonbon 'candy', Fenster 'window', Sofa 'sofa', Tee 'tea'.

Again, no single suffix can be productive according to the statistics in (32). Since the total number of [-fem] nouns is 292, a productive rule can have no more than $\theta_{292} = 51$ exceptions; again, none of the suffixes pass the test.

One outcome is to lexically list every [-fem] noun with its suffix, but the child is not quite ready to give up. For that matter, nor will the Tolerance Principle allow us to stop here. Within the finer subclasses of nouns, productive generalizations are to be expected.

4.4.2.2 Phonological Regularities

The most immediate step is to consider the role of phonology. As seen in the quote from Wiese 1999 earlier (p. 124), [-fem] nouns with a "reduced final syllable" (RFS, a schwa followed by *l/r/n*) tend to take the null suffix -ø; see also see Dressler 1999 and Wunderlich 1999 for similar suggestions. This is strongly confirmed in our data. There are 83 [-fem] nouns with a reduced final syllable (+RFS):

(33) a. 77 take the null suffix -ø.

b. 3 take -*n*: Bruder 'brother', Rätsel 'puzzle', Zahnstocher 'toothpick'.

c. 3 take -*s*: Fenster 'window', Garten 'garden', Pullover 'sweater'.

The 6 exceptions are nowhere near the tolerance threshold ($\theta_{83} = 18$) and so I predict a productive rule:

(34) For the [-fem] nouns: If [+RFS] THEN -ø

As a matter of fact, rule (34) removes virtually all the -ø suffixed nouns from consideration.

An additional phonological consideration concerns nouns that end in a schwa. It has been observed that nouns ending in a schwa add -*n* in plurals (e.g., Bittner 2000a). This seems to be a nearly categorical generalization regardless of gender. Table 4.7 presents the type frequencies of the nouns that end in a schwa and their suffixation choices from the CELEX Corpus as reported by Bartke et al. (2005, 49).

Table 4.7
Type frequencies of nouns with final schwa and their suffixes based on the CELEX Corpus

Gender	-(e)n	-s	-ø (+Umlaut)
+fem ($N = 903$)	99.9% (902)		0.02% (1)
-fem ($N = 97$)	86.5% (84)	2.1% (2)	11.3% (11)

No matter how one evaluates the pattern in Table 4.7, the -*n* suffix is clearly productive for nouns with a final schwa. It is almost exceptionless for [+fem] nouns. For the [-fem] nouns, $N = 97$ can tolerate $\theta_{97} = 21$ exceptions when there are only 13. In our child-directed corpus, every noun that ends in a schwa adds -*n*, including the 11 [-fem] nouns list below.

(35) Affe 'monkey', Auge 'eye', Drache 'dragon', Getreide 'grain, cereal', Hase 'rabbit', Löwe 'lion', Name 'name', Ochse 'ox', Orthopäde 'orthopedist', Rabe 'raven', Seite 'side, page'

Thus, we have two phonological processes that target a "light" final syllable — /ə/ in one, /əl/, /ər/ and /ən/ in the other — and productively attach a suffix (-*n* and -ø respectively). It is possible that they can be further unified. Wiese (1996, 106), for instance, treats both processes as a reflex of satisfying the prosodic constraint that plural nouns must end in a bisyllabic foot, with

4.4 The Horrors of German: Exceptions that Force the Rules

the second syllable containing a schwa. For our purposes, these rules have the effect of removing 94 competitors—that is, 83 in (34) and 11 in (35)—that would otherwise be exceptions to the -*e* suffix.[22]

Now child learners have only 198 [-fem] nouns to worry about. The -*e* suffix is overwhelmingly preponderant because it accounts for 143 or 72% of the nouns. But the 55 exceptions still fail to support a productive rule ($\theta_{198} = 37$). There must be additional subregularities: further partitioning is necessary.

4.4.2.3 The Masculine Nouns

Consider now the division between masculine ([+masc]) and neuter ([+neut]) nouns. For both classes, the -*e* suffix has been suggested as the default by previous scholars (e.g., Wiese 1996, 138 and Laaha et al. 2006).

The [+masc] class is now completely straightforward. After the nouns that take phonologically predictable suffixes are accounted for, there are 129 masculine nouns left:

(36) a. 107 take -*e*.
  b. 8 take -*en*: Dorn 'thorn', Fleck 'spot', Held 'hero', Mensch 'human', Papagei 'parrot', Schmerz 'pain', Teddybär 'Teddy bear', Vormittag 'morning'.
  c. 6 take -*er* all with Umlaut: Gott 'god', Mund 'mouth', Ort 'place', Rand 'edge', Strauch 'shrubbery', Wald 'forest'
  d. 4 take -ø: Nachbar 'neighbor', Straßenrand 'roadside', Takt 'pace, time', Tintenfisch 'squid, inkfish'.
  e. 4 take -*s*: Liebling 'loved one', Park 'park', Tee 'tea', Zoo 'zoo'.

Some of the non-*e*-taking exceptions (e.g., *Tintenfisch*, literally 'inkfish') are probably compounds and should not been have counted beyond the root. But even if the child learners have not identified the compound structure of these words, there are still not enough exceptions to derail the productivity of -*e*. A total of 129 items can tolerate up to $\theta_{129} = 26$ exceptions and there are at most 22 in (36). Thus, we conclude that the default suffix for [+masc] nouns is -*e*.

---

[22] The near absence of the schwa ending has been shown to be one of the phonetic cues for gender acquisition by children (Mills 1986; Szagun 2004).

4.4.2.4 The Neuter Nouns

Finally, consider the [+neut] nouns. Again, we move those with phonologically predictable suffixes out of the way and consider the remaining 69. This turns out to be a very mixed bag: 36 take -*e*, 24 take -*er* (some followed by Umlaut), 4 take -*(e)n*, and 5 take -*s*. The predominant suffix -*e*, conjectured to be the default (Wiese 1996), is still quite far from becoming the default: $\theta_{69} = 16$, more than the actual number of exceptions (33) to the -*e* suffix.

Lexicalization? Not so fast. There is yet another regularity that will handle most of the -*er* nouns, which are the biggest group of competitors for -*e*. First, for monosyllabic neuter nouns that contain the back vowels /a, o, u/, a strong tendency is to add -*er* followed by Umlaut:

(37) Monosyllabic neuter nouns that with the back vowels /a, o, u/:

    a. 16 take -*er* followed by Umlaut: Band 'pool', Blatt 'leaf', Dorf 'village', Glas 'glass', Haus 'house' Holz 'wood', Kalb 'calf', Kraut 'plant', Land 'land, country', Maul 'jaw', Loch 'hole', Schloss 'castle', Schloss 'lock', Tuch 'cloth', Volk 'people', Unkraut 'weed'

    b. 6 take -*e*: Haar 'hair', Jahr 'year', Paar 'pair', Pfund 'half kilo', Rohr 'pipe', Tor 'door, goal'

The distribution in (37) supports a productive pattern: $N = 22$ can tolerate seven (θ_{22}) exceptions, and the six -*er*-suffixed nouns are manageable.

So the -*er* plus Umlaut rule in (37) shaves off some additional exceptions, and the remaining 46 are presented below:

(38) a. 30 take -*e*.

    b. 7 take -*er*: Bild 'photo', Brett 'cutting board', Feld 'field', Gesicht 'face', Lied 'song', Nest 'nest', Spiegelei 'fried egg'.

    c. 4 take -*en*: Aquarium 'aquarium', Datum 'date', Hemd 'shirt', Ohr 'ear'

    d. 5 take -*s*: Bonbon 'candy', Kilo 'kilo', Restaurant 'restaurant', Sofa 'sofa', Video 'video'

As far as the numbers go in (38), the -*e* suffix is just short of productivity: it has 16 exceptions, whereas $\theta_{46} = 12$. Several subregularities can still be identified. The Latinate nouns with the -*um* ending in (38c) tend to add -*en*, as noted by many German grammar guides. And the -*s* suffixed nouns in (38d)

4.4 The Horrors of German: Exceptions that Force the Rules

are clearly loanwords — none need translation! — which appear to fall under the -s suffix, as I discuss shortly. Moreover, I believe that the productivity for -e would be assured if we enlarged the vocabulary to include just a few more neuter nouns. The number of -e suffixed nouns will keep increasing, but sooner or later the lexically idiosyncratic -er suffixed exceptions will run out. For instance, I searched in a list of the 1,000 most frequent German noun plurals. There are 222 neuter nouns, as opposed to 106 in our child-directed corpus. The numbers of -er taking nouns in the two lists are 12 and 8 respectively, a very modest difference, but the numbers of -e taking nouns almost double (72 vs. 35). Thus, after factoring out the phonologically predictable suffixes, there is every reason to believe that -e can be identified as the productive default for neuter nouns.

4.4.3 German Plurals: A Recap

Let's summarize the key features of the German noun plural system.

As Table 4.6 makes clear, the Tolerance Principle predicts that the German plural system cannot have only one productive suffix. My analysis has been driven by the need to discover productive processes among the "irregular" nouns and arrived at a conclusion very similar to previous proposals by German morphophonologists. Specifically, the realization of the plural morpheme (PL) follows these rules:

(39)   a. PL → -(e)n / [+fem]
      b. PL → ø /[+RFS] __ #
      c. PL → -n /ə __ #
      d. PL → -er[+Umlaut] for monosyllabic neuter nouns with back vowels /a, o, u/
      e. PL → -e
      f. PL → -s

The German plural system is an example of nested rules par excellence; see (28) as well as the discussion in Section 3.5.4. The computation of the rules in (39) is sequential in the order specified. A noun marked with [+fem] will not make it past (39a), and the rules after it are devoted to nonfeminine nouns. I leave open the question of whether the implementation should be done with extrinsic ordering or through the Elsewhere Condition with appropriate feature specifications; see Halle and Marantz 1993 and Noyer 1992 for alternative views. I also leave open the ordering of (39b-c): both are phonological in nature and may be unified (Wiese 1996).[23] Finally, virtually every rule listed here has exceptions, but they remain productive because the exceptions do not exceed the Tolerance threshold and can be lexicalized for suffixation.

Algorithmically, we first check if a noun is feminine: rule (39a). If so, the noun is searched against a list of exceptions that do not take *-(e)n*; otherwise the *-(e)n* suffix is applied.

If the noun is not feminine, rules (39b-c) take effect. They are phonological in nature, perhaps reflecting the general phonotactic properties of German (see Wiese 1996 and the discussion above). Nouns that end in a reduced syllable (RFS: *el/er/en*) add the null suffix -ø, and nouns that end in a schwa add -*n*. Again, both rules are productive, but their lexically listed exceptions are evaluated first following the Elsewhere Condition.

Next is rule (39d), which applies to monosyllabic neuter nouns that contain the back vowels /a, o, u/. The *-er* suffix is used for these nouns, which seems to invariably trigger Umlaut. Again, a list of exceptions that satisfy the structural properties of (39d) but do not take *-er* must be lexicalized.

Rule (39e) specifies the default suffix *-e* for [-fem] nouns. My analysis shows that after the phonological properties in rules (39b-d) are accounted for, the *-e* is the productive suffix for both [+masc] and [+neut] nouns; I leave open the question of whether they should be unified as the [-fem] class or kept disjunctive.

Finally, we have (39f). There are thirteen nouns in all that take *-s* in the corpus, a mixture of all three genders with no clear statistical tendency whatever. Thus, unlike rules (39a-e), which all refer to gender, the *-s* suffix places no restrictions on gender, and thus is most general, making it the default rule. I do not have sufficient information from the child-directed German corpus to determine whether this rule has additional regularities. For examples, it has been suggested that the *-s* suffix is specialized for loanwords; Wunderlich

---

23 It is possible, as suggested by Abby Cohn (personal communication), that all nouns with a final schwa, regardless of gender, immediately attach *-n* for plurals, prior to the consideration of grammatical gender ([±fem]). This will lead to a slightly different order of rule application: (39c) would be moved to the front of the list in (39). The actual organization of the rules by German-learning children may depend on the strength of the phonological cues vs. the grammatical gender cue in their vocabulary. Moreover, the two alternative representations make different behavioral predictions if we, as before, take rules such as (39) as an algorithmic model of language use. For instance, if the rule for nouns ending in a schwa regardless of gender is placed at the top of the list, then we expect feminine nouns that end in a schwa to be pluralized faster than those that do not, after controlling for stem and surface frequencies. If no difference is found, then the [+fem] rule applies first as in (39).

4.4 The Horrors of German: Exceptions that Force the Rules

1999, for instance, postulates a [+foreign] feature and attributes the suffix to English or French borrowing from the nineteenth century, but see Wiese 1996, 137-138 for an opposing view. It is also recognized that proper names (e.g., *Lehmann-Lehmanns*), onomatopoetics (*Wauwau-Wauwaus* 'dog'), abbreviations (*Prof* for *Professor*), and so on take *-s*; see Marcus et al. 1995. Presumably these properties will be learned later in life. Special phonological characters may be at play — for instance, long vowels at the end of a noun such as those in (38d), which are quite unGerman, may be telltale signs for the application of the *-s*.

Let's consider the type of plural errors that German-learning children make. Since all the rules in (39) have exceptions but are nevertheless productive, each would function like the past tense of English: the lexically idiosyncratic exceptions may be regularized once the productivity of the rule is established. That is, a [+fem] noun marked as an exception to the *-(e)n* may occasionally take the suffix instead; these are common errors in German acquisition (e.g., Gawlitzek-Maiwald 1994), just like the irregular verbs occasionally taking *-d* in child English. By the same notion, there are nouns that are exceptions to the morphophonological rules (i.e., (39b) and (39e)). For instance, the word *Hunde* 'dogs', already in the plural, is sometimes suffixed with an additional *-n* (*Hunden*; Szagun 2001), perhaps by virtue of ending in a schwa. Further down the list, we have the default suffix *-e* for [-fem] nouns but there are again exceptions. For example, the neuter noun *Herz* 'heart' obviously does not meet the phonological conditions in (39b), (39e), or (39d) but does not take *-e* either: its plural form is *Herz-en* and thus must be lexically marked for *-en*. Yet young children sometimes produce *Herze* for the plural form (Szagun 2001). Of course, our discussion presupposes that the children have learned the gender of the nouns correctly: if a [+fem] noun is mislearned as [+masc] — children's acquisition of gender marking is excellent but not perfect (Szagun 2004), — then they may deviate from the target form even further. For instance, some children produce *Fischen* instead of *Fische* for the plural of *Fisch* 'fish', a masculine noun (Gawlitzek-Maiwald 1994, 248). The analysis proposed in (39) asserts that this type of error must be the result of marking *Fisch* as feminine, since there is no productive rule that would attach *-en* for [-fem] items that do not end in a schwa. Finally, the *-s* suffix is overused when any of the prior rules fail to apply, and possibly for nouns whose gender the child is not sure of — the *-s* is unmarked for gender and is thus in principle consistent with any word. Presumably this could account for why loanwords often take *-s*: if the gender of a new word is yet to be firmly conventionalized, then none of the four suffixes is applicable because all make specific reference to gender.

To reiterate my central point; the final analysis of the German plural system is initially motivated by the total absence of any productive suffix in the input data (Table 4.6). The learning process follows the recursive application of the Tolerance Principle, as we — or rather, German-learning children — are led to discover productive regularities within the subdivisions of the nouns, which are specified by well-motivated gender and phonological features. Each rule identified in (39) deals with a subset of the nouns: few are exceptionless but none exceeds the threshold of Tolerance. Each rule, then, cuts down the total number of nouns to be accounted for so that the default suffixes — *-(e)n* for [+fem] and -e for [-fem] — may achieve productivity for significant subsets of the nouns. The complexity of the German plural system, and the convergence between my numerically driven analysis and previous theoretical treatments, provides evidence for the Tolerance Principle as a key component of a theory of morphological learning.

So there is order in the apparent chaos of German plurals after all. But the horrors of the German language haven't exactly gone away: the child now just has to learn the marking of gender.

****

This has been an exhausting chapter, but the details are necessary in order to establish the credibility of the Tolerance Principle when engaging with the complex reality of natural languages. I hope that the explanations have been sufficiently mechanical, perhaps nauseatingly so, such that the Tolerance Principle may eventually be implemented in a general learning model with support from additional components of the grammar (phonology, syntax, semantics, etc.).

The main theme of the case studies can be summarized succinctly: the child language learner is quite single-minded in the pursuit of linguistic productivity. Lexicalized listing is always an option: every word is an instance of Sausserean arbitrariness, and children are extremely proficient word learners. But when it comes to the generative component of languages — forming past tense, pluralizing nouns, assigning stress, deriving new categories of words — children seem to do their very best to avoid brute-force memorization. When a set of data fails to yield regularity, the next move is to partition the data into subsets, along some appropriate linguistic dimension, such that productive regularities

4.4 The Horrors of German: Exceptions that Force the Rules

may arise within. Qualitative changes in the grammar follow from the quantitative accumulation of linguistic experience.

These studies, I believe, are of general interest to linguists confronted with a body of heterogeneous data or different degrees of granularity: the core vs. periphery problem again. The existence of exceptions to varying degrees does not mean the absence of an overarching grammatical system such that everyone must be treated on a kline of productivity (Goldberg 2003; Hay and Baayen 2005; Jackendoff 2007; McClelland and Bybee 2007). To the extent that all grammars leak, the Tolerance Principle provides a quantitatively based criterion for treating data heterogeneity as a single system with exceptions or as multiple parallel systems that have independent grammars within, which may also have exceptions.[24] If this is correct, then there is no a priori reason to prefer either approach; the correct theory will be decided by the numbers in the data. I further speculate that the Tolerance Principle may play an important role in language acquisition in a multilingual learning process: the problem of distinguishing several languages is formally equivalent to the problem of distinguish sub-regularities in a single language. A handful of strange words or visiting relatives with a silly accent will not disrupt the acquisition of a single linguistic system, just as a few irregular verbs do not undermine the regular "add -*d*" rule. But if the learner's environment consists of significant quantities of multilingual data, then no single language is likely to tolerate the others as exceptions. The learner will be compelled to partition the input into distinct subsystems and develop independent grammars for each, much like the English stress and German noun systems studied here.

The search for productivity may grind to a halt, when all plausible avenues have been explored. Recall the learning procedure in Figure 1.1: What if none of the rules satisfies the Tolerance Principle? At some point, children must give up since no one learns forever. And give up they do: the irregular past-tense rules in English all succumb to cumulative exceptions over time (section 4.1), relegating them to lexicalization. Exceptions, when recognized as such, can be harmlessly lexicalized away but as we will see in chapter 5, they also have the potential to upend the grammar.

---

24 See, for instance, the debate between the indexed constraint approach (Itô and Mester 2003; McCarthy and Prince 1995) and the cophonology approach (Anttila 2002; Inkelas and Zoll 2005) in the analysis of morphologically conditioned phonological processes.

5 When Language Fails

The hallmark of language is discrete infinity: as reviewed in previous chapters, even young children are capable of extending linguistic patterns to novel contexts. It is therefore puzzling to find dark and dusty corners of language where productivity unexpectedly fails.

In a classic paper, Halle (1973) draws attention to several types of morphological "gaps." For instance, there are about seventy verbs in Russian that lack an acceptable first-person singular non-past form (data from Halle 1973, 7, and Sims 2006).

(1) *muč/*muščuʼ 'I stir up'
    *očučusʼ/*očuščusʼ 'I find myself'
    *pobežu/*pobeždu 'I win'
    *erunžu/*erunždu 'I behave foolishly'
    *lažu/*lažd'u 'I climb'

There is nothing aberrant about the form or meaning of these words that could plausibly account for their illicit status. Yet native Russian speakers regard them as ill-formed.

Arbitrary lexical gaps[1] like those in (1) are in fact quite common in the world's languages; see Baerman and Corbett (2010), Fanselow and Féry (2002), and Rice and Blaho (2009) for surveys. (And as we will see in chapter 6, the acquisition of negative constraints in syntax poses very similar challenges and receives very similar solutions.) Even the relatively impoverished morphology of English contains gaps—for example, the past participle form of *stride* is famously lacking, with few speakers accepting either *stridden* or *strided* (see among others Pullum and Wilson 1977, 770; Pinker 1999, 136f.). Inflectional gaps were first noted in ancient Roman times (Neue 1866, 503f.), but they have not received sufficient attention in contemporary theories of word formation. For instance, gaps are unexpected under any competition-based approach that assumes the existence of a default or winning form, which includes many prominent theories of language and language acquisition such as Optimality Theory (Prince and Smolensky 2004), Distributed Morphology (Halle and Marantz 1993), Paradigm Function Morphology (Stump 2001), Network Morphlogy (Brown and Hippisley 2012), and the dual-route model of the lexicon (Clahsen 1999; Pinker 1999).

---

1 I adopt this traditional term (Halle 1973; Hetzron 1975) as opposed to the more recent use of paradigmatic gaps to refer to the phenomenon as in (1). I see no need to commit to theoretical notions such as the paradigm in my treatment of gaps.

At this point, it should be noted that gaps can be "productive," in the sense that ineffability is not necessarily limited to a finite number of lexical items in the language but is sometimes extended to novel words. In a study of the Russian gaps, Baronian and Kulinich (2012, 92) note that recent lexical entries into the language may be spontaneously gapped in the 1SG nonpast. The authors quote a blogger's ambivalence toward the inflection of the verb *friendit'* 'to friend': "Friending policies: I accept new friends, but I don't make friends [frend'u] (or frenžu?) only for reciprocity. However, I make friends [frend'u](-žu) whom [I find] interesting and pleasant to read." When presented with neological verbs borrowed from English that have similar stems as the defective verbs, such as *(ot-)routit'* 'to route' and *apgrejdit'* 'to upgrade', Russian speakers in Baronian and Kulinich's study produced responses very similar to the existing defective verbs—and markedly different from nondefective verbs, even those with very low frequencies. Knowledge about gaps, then, is not restricted to the existing words but reflects an abstract property of the language projected from the linguistic input. This invites a unified approach to gaps as well as productive processes in word formation.

I show that the Tolerance Principle provides a predictive account of when gaps arise, or more precisely, when productivity breaks down. The present chapter has two parts. First, I use the Tolerance Principle to find gaps in several well-known cases in English, Polish, Spanish, as well as Russian in Halle's original discussion. Second, I propose that the Tolerance Principle holds the key to understanding how productivity rises and falls, which is most clearly seen during the course of language change. Since productivity is a numerical matter between rules and exceptions (N and e), I present a detailed study of the so-called Dative Sickness in the morphosyntactic system of Icelandic, where we show that the attested change is predictable on the basis of lexical statistics.

5.1 Finding Gaps

Halle's (1973) original study of gaps proposes two analyses (which he regards as equivalent). In his first, and better-known, proposal, the words in (1) are in fact generated (or at least could be generated) by the word-formation component, but their actual use is blocked by the feature [−Lexical Insertion]. This idea is key to many subsequent treatments of defective gaps under the principle of "lexical conservatism" (e.g., Pertsova 2005; Steriade 1997): do not use a form unless it is explicitly attested. Lexical conservatism is also implicit in

5.1 Finding Gaps

some analyses of gaps in Optimality Theory (e.g., Prince and Smolensky 2004; Raffelsiefen 1996, 1999, 2004; Rebrus and Törkenczy 2009; Rice 2005; Wolf and McCarthy 2009), which posit that ineffability occurs when a phonologically null candidate (the "Null Parse") is optimal. This requires a grammar in which markedness dominates faithfulness, a property of the "initial state" (Smolensky 1996). Distributed Morphology analysis approaches gaps in a similar fashion. Embick and Marantz (2008, 22), for instance, take forms such as *forwent to be generated by the grammar but somehow regarded as degraded by the speaker. But proposals along the lines of [−Lexical Insertion] have difficulty reconciling the presence of gaps with the unbounded creativity of word formation: children do pass the Wug test, and when new words join the English lexicon, their inflectional forms are instantly available. A principled division between gaps and productivity must be drawn; invoking lexically arbitrary features or constraints only restates the problem.

Alternatively, Halle (1973 note 1) suggests that only those items undergoing what he calls "nonproductive" word-formation rules are eligible for the feature [−Lexical Insertion]; if the generated form is in fact attested in the language, it will bear the feature [+Lexical Insertion], allowing it to be used. Hetzron (1975), while arguing against Halle's better-known proposal of lexical conservatism, makes essentially the same suggestion. Rules are either productive or lexicalized, and gaps arise in the unproductive corners of the grammar. His conception of gaps can be strongly identified with the Elsewhere Condition, a critical component of the present theory:

> The speaker must use ready-made material only for 'exceptional' forms, while everywhere else he could very well 'invoke the word formation component'. Technically, this can be represented by a disjunctive set of rules where idiosyncratic or 'exceptional' formations are listed with as much explicitness as necessary, while the general word formation rules would appear afterward, with the power to apply 'to the rest' (871).

Subsequent proposals have adopted similar positions (Albright 2009; Anderson 2010; Baronian 2005; Hudson 2000; Maiden and O'Neill 2010; Pullum and Wilson 1977; Sims 2006), attributing nonproductivity to conflicting structural and/or statistical factors in phonology or word formation. Under this view, gaps arise when there is no productive/default rule to fill in the slot. In such cases, the speaker requires positive evidence for every slot; without such evidence, ineffability occurs. These accounts differ considerably in implementation, but none — except Albright 2009; see section 5.1.2 — provides an account of how and when conflicting forces in word formation produce gaps.

Under the Tolerance Principle, mere majority status does not entail productivity. Only a filibuster-proof supermajority will do, because the sublinear growth of the function $\theta_N = N/\ln N$ limits the number of exceptions to a very small proportion (Table 3.1). When this threshold is exceeded, no productive rule can be identified for the context. The emergence of gaps, however, is not only a matter of statistics. Recall the case of German noun plurals (section 4.4): there are five suffixes but none comes anywhere near the statistical majority required by the Tolerance Principle. In that case, however, there are well-motivated grammatical and phonological factors that partition the nouns into subclasses, and within each productive suffixes can be identified. To lexical gaps to arise, then, these structural conditioning factors cannot be present, or in any case must be beyond the child learner's ability to detect. Only then will the absence of a sufficiently dominant alternation force the learner to lexicalize the attested forms, and to have nothing to offer for unattested forms. More specifically, gaps arise if the following numerical condition is met:

(2) Conditions on gaps

Let there be S alternations, each affecting N_i lexical items ($1 \leq i \leq S$), and $\sum_i N_i = N$. Gaps arise if and only:

$$\forall i, 1 \leq i \leq S, \sum_{j \neq i} N_j > \theta_N$$

None of the alternations (N_i) in N are sufficiently numerous to tolerate the rest ($\sum_{j \neq i} N_j$) as exceptions.

I provide four case studies to validate this hypothesis. Note that my goal in each case is to predict *contexts* in which gaps arise, rather than to predict exactly which item is ineffable (e.g., Albright 2003; Hudson 2000) or how the speaker may compensate for such gaps (by finding a paraphrase, for instance). I adopt this more modest goal due to my suspicion that gaps are generally arbitrary and are accidents of history (see, e.g., Baerman 2008 for an extensive discussion of the defective verbs in Russian) and the failure of models such as Albright 2003 to accurately predict gaps (see section 5.1.2). Seeking deep regularities within contingencies misses the point.

5.1 Finding Gaps

5.1.1 Stride, Strode, *Stridden

I start the empirical study of gaps with a perennial puzzle. It has long been noted that certain English irregular verbs lack a past participle form. For example, English speakers know the past tense of *stride* is *strode* but few are willing to accept either **stridden* or **strode*, never mind **strided*, as the past participle. The presence of the gap is readily confirmed by corpus statistics; see Baronian 2005 for a similar treatment. For instance, in the 450-million-word Corpus of Contemporary American English (COCA; Davies 2008), the past participle of *stride* appears only three times:

(3) a. Roxanne's unleashed herself from her laptop, while Angela has **stridden** into the country without her makeup and mirror. (*Shape* Magazine, 1999)

b. And he'd shattered the screen of his monitor with one kick of his oaken legs, hauled open the steel door normally used only by Ted, and **stridden** into the Grim Reaper personally. (Science fiction: *Bulldog Drummond and the Grim Reaper* (1996) by Michael Coney)

c. Teenaged me would have **strode** up to that bank and plunged in — fully clothed and come-what-may. (*Backpacker* Magazine, 2011)

Of course, the very low frequency and even the absence of a linguistic form in some corpus does not mean it's unavailable or ungrammatical, especially given the Zipfian nature of linguistic distributions (chapter 2). But the statistical rarity of *strode/stridden* in COCA does strongly suggest that the gap is genuine, at least for most speakers of English. First, the absence of *stridden* or *strode* is surely not a problem of its semantics or phonology. Similar verbs such as *slide*, *ride*, and *abide* are not gapped, and *-en* is among the more systematic, though still unproductive, patterns for past participles. Second, *stride* is hardly a rare verb; if it were so, then the past particle form may also be expected to be rare in a corpus. Here I use the 100-million-word British National Corpus (BNC; BNC Consortium 1995), which, unlike COCA, contains both word frequencies and their parts of speech. There are many verbs, including numerous regulars and a nonnegligible number of irregulars that are less frequent than *stride* in the infinitive form. Counting only morphologically simplex irregulars (thus excluding words such as *overhear*, which is derived from *hear*), these include *bust*, *slay*, *sling*, *stink*, *shit*, *smite*, etc., all of which have uncontroversially attested past participles. Finally, and importantly, the past-tense form of *strode* is in fact relatively common, more frequent than *fled*, *swore*, *dug*, and numerous others. An English speaker will have ample evidence that *stride* is an irregular verb.[2] This is a crucial point because had *strode* been rare, the speaker may have regarded *stride* as a regular verb and *strided* would have been the past participle. For instance, *chide* traditionally was inflected as *chid* and *chidden* for past tense and past participle but these forms are now virtually absent. All seventeen occurrences of *chid* in COCA are spelling variants, or possibly typographic errors, of *child* and it does not exist at all in the BNC; *chidden* is absent in both corpora. In all likelihood, *chide* is now a regular verb (*chide-chided-chided*). These frequency considerations establish that the *stridden* gap is genuine, but they also raise an interesting point regarding frequency and grammaticality. In a sufficiently large corpus such as COCA, the absence or very low frequency of an inflected form is conspicuous and can be viewed as evidence for its ungrammaticality or unavailability — *only* if the form concerns an irregular item (such as *stride*). The regulars, which can always add new members, are not hindered by the absence of inflected forms. This suggests that the irregularity of a word, which is indicated by the nonregular past tense form of -*strode*, is a necessary condition for inducing the *stridden-style gaps. Furthermore, the statistical rarity or even absence is not sufficient for gaps or ungrammaticality in general: there is considerable confusion on this matter, which I discuss in chapter 6 in the context of indirect negative evidence in language acquisition.

Let's consider the range of inflectional processes that English speakers may have at their disposal when the past participle of *stride* is called for. The most promising pattern to pursue is the syncretism between past and past participle. Learners may have observed a great majority of irregular verbs with identical past tense and past participle forms and decide that it is a general pattern that can be extended to verbs for which the past participle form has not been observed. Here I take no position whether this is a plausible grammatical analysis for the derivation of the past participle. For instance, Embick and Halle (2005) analyze the past participle as derived from the root/stem, whereas Albright (2006) considers the past participle to be derived from the past tense. My main purpose is to show that the gap for *stride* is due to the absence of any generalizable pattern among the English irregular verbs, and I can only do so by considering all conceivable patterns of generalization and showing none is

---

2 Although two instances of *strided* as past tense did appear in COCA.

5.1 Finding Gaps

up to par. Since the derivation of past participles from the past tense has been proposed before, it falls into the realm of conceivable generalizations.

But the syncretic rule does not fare well enough. There are 95 irregular verbs in our 5-million-word child-directed corpus of English, of which 52 have identical past and past participle forms. In the BNC, there are 226 irregular verbs that appeared in the past tense; of these, 141 have syncretic past and past participle forms. Since some of the 226 irregulars in the BNC are very obscure, let's consider the set that is more likely to represent an English speaker's vocabulary: an irregular verb is included if its past tense appears at least once per 10 million. This restriction obtains 184 irregular verbs, of which 108 are syncretic between past and past participle. Restricting the frequency even further to once per million, we obtain 131 irregular verbs, of which 72 are syncretic between past and past participle. The results are summarized in Table 5.1.

Table 5.1
The number of irregular verbs that are syncretic in the past and past participle in several lexical sets from the CHILDES and the BNC corpus. The BNC corpus has 100 million words, thus a frequency of ≥ 100 corresponds to at least once per million.

Corpus	N	Syncretic	Non-syncretic	θ_N
CHILDES	95	52 (55%)	43	21
BNC (freq$\geq$100)	131	72 (55%)	59	27
BNC (freq$\geq$10)	184	108 (59%)	76	35
BNC (all)	226	141 (62%)	85	42
BNC (freq$\geq$100, simplex)	115	69 (60%)	56	24
BNC (freq$\geq$10, simplex)	141	89 (63%)	52	28
BNC (all, simplex)	147	95 (65%)	52	29

In Table 5.1, I have also restricted the irregular verbs even further by considering only morphologically simplex (underived) verbs (i.e., excluding *offset, outgrow, withdraw, retake*, etc.). Some of these prefixes are clearly no longer compositional — *withdraw* is not clearly related to *draw* — but I wanted to give learners every opportunity to identify productive patterns because a smaller N in the Tolerance calculation ($\theta_N = N/\ln N$) allows for more exceptions (i.e., nonsyncretic forms). But as the results in Table 5.1 make clear, no matter how the list of irregular verbs is constructed, the "syncretic rule" between the past and past participle never approaches the numerical advantage requisite for productive extendability, even though it is consistently the majority pattern (55% to 65%). There are simply too many irregular verbs for which the past and past participle are not the same (e.g., *break-broke-broken*), which are the exceptions to syncretism.[3]

This rules out *strode* as a past participle for *stride*, but what about *stridden* along the lines of *ride-rode-ridden*? Or indeed, *stroden* along the lines of *broke-broken*? More applications of the Tolerance Principle. Consider how learners could conceivably derive *stridden* or *strode*. We need to evaluate all irregular verbs that fit the structural description of the stem (N) and see how many of them actually follow a specific pattern. On the list of BNC simplex irregular verbs, the following fourteen stems contain the diphthong /aɪ/ followed by a single consonant, which is the most conservative generalization for the relevant class that involves *stride*.

(4) a. ai→o~en: arise~arose~arisen, drive, rise, ride, strive, write (6)

b. ai→ɪ~en: bite~bit~bitten (1)

c. ai→ʌ~en: strike~struck~striken (1)

d. ai→a (syncretic past and past participle): fight (1)

e. ai→ɪ (syncretic past and past participle): hide, light, slide (3)

f. gaps: dive (?), strive (2)

The route of *ride-rode-ridden* simply has too many exceptions to be available to verbs such as *dive* and *strive*.

Consider then the route of *broke-broken* where the past participle simply affixes *-en* to the past-tense form; again, I put aside the theoretical formulation of these analyses but focus on the detectable pattern in the data instead. There are twenty irregular verbs from the BNC simplex irregular verbs that take *-en* in past participles:

(5) a. participle = past + [n̩]: awake, bid (?), bite, break, choose, freeze, hide, speak, steal, wake (10)

b. participle = stem + [n̩]: arise, eat, drive, fall, give, ride, rise, shake, strike, take (10)

---

[3] In some dialects of American English, speakers in certain contexts routinely produce the simple past form of irregular verbs when a past participle is called for (e.g., *I have ate the cake*). In such linguistic environments, the acquisition data would presumably show a sufficiently high level, and thus productive use, of syncretism between the simple past and the past participle. These speakers will therefore not have a past-participle gap for *stride*. Instead, they produce expressions such as *I have strode home*, as reported in an informal survey of Midwestern American English speakers (Kyle Latack, personal communication).

5.1 Finding Gaps   147

The extension along the lines of *broke-broken* has no chance of success either.

In conclusion, no conceivable route for past-participle formation can survive the Tolerance test in any reasonable sample of English words. Therefore, once learners recognize a verb is irregular—for example, on hearing *strode*, which does not bear the regular *-ed* suffix—they will need positive evidence to complete the inflectional paradigm. The absence of an attested form, and the absence of any productive rules, create gaps when the past participle of *stride* is needed.

5.1.2 Stem Alternations in Spanish

The inflection of many Spanish verbs exhibits stem alternations targeting the final vowel of the verb; see section 2.3.3 for an overview of Spanish inflection and its acquisition.[4] One of the most prominent alternations is diphthongization, which takes place in both inflectional and derivational morphology (Eddington 1996; Harris 1969, 1977; Malkiel 1966). Many verbs that have a stem-final [e] or [o] in unstressed position change it to *ie* [je] and *ue* [we] in stressed positions (from Albright, Andrader, and Hayes 2001).

(6) [sentámos] ∼ [s**jé**nto] 'we/I sit'
     [tendémos] ∼ [t**jé**ndo] 'we/I stretch'
     [podémos] ∼ [p**wé**do] 'we/I can'
     [kontámos] ∼ [k**wé**nto] 'we/I count'

The diphthongization process can be traced back to historical developments in Spanish, but the current set of verbs that participate in diphthongization is lexically arbitrary. Consider the contrast between (6) and (7), where the vowels surface as [e] and [o] in both stressed and unstressed forms:

(7) [rentámos] ∼ [rénto] 'we/I rent'
     [bendémos] ∼ [béndo] 'we/I sell'
     [podámos] ∼ [pódo] 'we/I prune'
     [montámos] ∼ [mónto] 'we/I mount'

Alternating verbs are a statistical minority in Spanish and can be found in all three conjugations. However, there are a number of third-conjugation verbs

---

4 This section reports joint work with Kyle Gorman, for which I am very grateful.

that simply do not permit any form of stem alternation and thus are unexpectedly gapped. Table 5.2 provides a comparison across the paradigm for the gapped *abolir* 'abolish' and the nongapped *dormir* 'sleep'.

Table 5.2
Partial paradigms for the verb *abolir*, which is gapped, as compared to the nongapped *dormir*. (After Maiden and O'Neill 2010)

abolir 'to abolish':

Pres. indic.	*	*	*	abolimos	abolís	*
Pres. subj.	*	*	*	*	*	*

dormir 'to sleep':

Pres. indic.	duermo	duermes	duerme	dormímos	dormís	duermen
Pres. subj.	duerma	duermas	duerma	durmamos	durmais	duerman

In addition to diphthongization, third-conjugation verbs participate in another type of stem alternation referred to as "raising" on the diachronic grounds that it historically involved the raising of the mid-vowel under assimilatory pressure (metaphony; see Malkiel 1966 and Penny 2002). For instance, the verb *pedir* 'to ask' in the present tense surfaces as *pido* in the first-person singular but *pedimos* in the first-person plural. Raising only appears in the third conjugation but has been argued to be lexically specific (Harris 1969). The language acquisition evidence is consistent with this analysis. As reviewed in section 2.3 (Clahsen, Aveledo, and Roca 2002; Mayol 2007), Spanish-learning children never mistakenly raise the vowel in the inflection of non-vowel-raising verbs, which suggests the absence of productivity (i.e., lexical specificity).

To identify verbs that exhibit this pattern of defectivity, I followed Sims's (2006) methods in her identification of gapped inflections in Modern Greek and Russian. I surveyed three Spanish dictionaries that provide information on defective paradigms (Butt and Benjamin 1988, Mateo and Rojo Sastre 1995, Real Academia Española 1992). I list all verbs which are defective according to at least two of the three sources.

(8) abolir 'to abolish'
    agredir 'to assault'
    aguerrir 'to harden'
    arrecirse 'to freeze'
    aterirse 'to freeze'
    colorir 'to color, dye'

5.1 Finding Gaps

despavorir 'to fear'
empedernir 'to harden'
preterir 'to ignore'
tra(n)sgredir 'to transgress'

I show that inflectional gaps in Spanish are treated the same way as in the missing past participle of *stride*. The language presents multiple possibilities for stem alternation; none, however, is sufficiently numerous so as to tolerate the rest as exceptions. Table 5.3 reports the frequency counts of candidate verb stems (namely, those in which the final stem vowel is mid) occurring at least once per million tokens in the LEXESP corpus (Sebastián et al. 2000). Many Spanish verb stems have multiple variants differing in the choice of the prefix; since prefixal variants on the same stem all exhibit the same stem changes (or "no change") under stress, verbs of the same stem are grouped together and counted only once.

Table 5.3
Distributions of stem changes in Spanish verbs and the productivity predictions of the Tolerance Principle

Conjugation	No change	Diphthongization	Raising	N	θ_N	Productive?
1st (-*a*-):	855	84	—	939	137	Yes
2nd (-*e*-):	115	21	—	136	28	Yes
3rd (-*i*-):	12	13	13	38	10	No

Table 5.3 shows that the distribution of diphthongization is quite different across the three conjugations. In the first and second conjugations, an overwhelming majority (91% and 85%) of "no-change" verbs do not participate in the alternation: both are safely above the tolerance threshold, and therefore both are the default process in these conjugations. These numerical considerations are consistent with findings from language acquisition reviewed in section 2.3.3. Children acquiring Spanish occasionally fail to apply stem changes to verbs of the first and second conjugations—that is, the overgeneralization of the productive "no- change" pattern. At the same time, they never overapply diphthongization to "no-change" verbs (Mayol 2007), strongly supporting the productive-vs.-unproductive asymmetry observed across languages. In a similar vein, Bybee and Pardo (1981) found that when presented with first-conjugation nonce verbs such as *bierca* and *duenta*, adult Spanish speakers

mostly produced *biercó* and *duentó* rather than *bercó* and *dontó* — that is, they treat the diphthong as part of the stem and did not apply diphthongization in verbal inflection. Thus, my prediction of "no change" as the default process in the first and second conjugation is confirmed.

Consider now the crucial case of the third conjugation, where gaps arise. According to the Tolerance-based approach, there should be no statistically dominant alternation. There are three alternations, including the trivial null alternation ("no change"), which add up to $N = 35$ verbs, which has the Tolerance threshold of $\theta_{35} = 10$. That is, if one of the three processes were to account for at least twenty-five verbs, then it can tolerate the remaining verbs as lexicalized exceptions. But clearly no process satisfies this numerical condition.[5] The absence of a productive alternation means that for every item in the third conjugation, learners must have positive evidence for its stem alternation because it is lexically arbitrary. And they will be at a loss for verbs that do not have attested forms, resulting in gaps.

For a comparison, consider one of the few formal approaches to the inflectional gaps in Spanish. In a series of papers, Albright and colleagues (Albright, Andrader, and Hayes 2001; Albright 2003) use a probabilistic rule induction model, the Minimal Generalization Learner (MGL; Albright and Hayes 2003), to predict which Spanish verbs undergo stem changes and which do not. In both studies, the model is trained on pairs of verb forms (e.g., *aprobar-apruebo* 'approve of', *pegar-pego* 'stick onto') with frequencies from the LEXESP corpus (Sebastián et al. 2000), inducing a list of stem-change rules. The model also generates a Confidence score associated with each rule, which is a function of the number of undergoers and exceptions to that rule. When two or more rules are in competition, the highest-Confidence rule is applied. Albright and colleagues (2001) report that this model accurately predicts human performance on a Spanish Wug test. In a later study, Albright (2003) asks speakers to use a seven-point Likert scale to rate their own "Uncertainty" about the correct inflection of various Spanish verbs. Uncertainty, averaged across participants, is negatively correlated with verb frequency and with the model's Confidence

---

5 Following Harris 1969 and Brame and Bordelois 1973, I assume that there is a unified rule of diphthongization that maps underlying /e/ to [je] and /o/ to [we] for the affected verbs; that is, the value of 13 in the "Diphthongization" column in Table 5.3 combines the two types of verbs. One might object to this treatment of diphthongization and prefer for it to be separated into two, more surface-based, alternations (Albright, Andrader, and Hayes 2001; Bybee and Pardo 1981; Eddington 1996). But a single process of diphthongization provides the most strenuous test of the Tolerance Principle. Diphthongization would stand a better chance of being productive if [je] and [we] verbs are unified.

5.1 Finding Gaps

score. From this, Albright concludes that Confidence and Uncertainty derive not just the contexts in which gaps occur, but also predict which possible words are ineffable.

I share these reseachers' intuition that the scope of a rule and the number of exceptions it faces play a role in determining productivity. But as Baronian (2005) points out, it is an open question whether speakers' subjective uncertainty or the MGL model's confidence score corresponds to the *presence* of inflectional gaps. Therefore, I investigated whether this model correctly classifies the defective verbs in (8). The MGL software was downloaded from the authors' website. Following Albright 2003, it was trained on pairs of candidate words (all nondefective third-conjugation verbs that are candidates for the diphthongization process) from the LEXESP lexical database (Sebastián et al. 2000). The defective verbs were "held out" for evaluation and the correct response should be the failure to generate any form, as would Spanish speakers. The model induces fourteen stem-change rules. For each of the third-conjugation verbs including the gapped ones, the final stem vowel (diphthongization or "no change") was generated using the highest Confidence rule. The results for the third-conjugation verbs are given in Figure 5.1. The model's Confidence score is plotted on the x axis and lemma frequency on the y axis; defective verbs are indicated with circles.

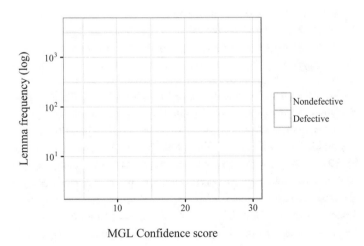

Figure 5.1
Spanish third-conjugation verbs and the predictions of the Minimal Generalization Learner

As can be seen from Figure 5.1, the presence of gaps in Spanish correlates with neither model Confidence scores nor lexical frequency. This echoes the observation earlier in the study of *stridden*: there is nothing conspicuous about the frequency of *stride* that makes it susceptible to gaps. Similarly, Sims (2006) finds that even very frequent words in Greek (e.g., *kopéla* 'girl') are gapped in the genitive plural, again dismissing any causal link between frequency and ineffability. It is also clear from Figure 5.1 that there is no threshold value for the MGL confidence score that would separate the verbs into gapped and ungapped classes; see Boyé and Cabredo Hofherr 2010 for similar conclusions when testing Albright's model on defective paradigms in French. The problem with the MGL model is that there will always be a rule that receives the highest Confidence score. Unless there is an independently motivated mechanism that relates the numerical score to ineffability, the MGL class of models is as problematic as the other competition-based theories that always predict an inflected form.

5.1.3 Defective Inflections in Russian

I now consider the best-known case of inflectional gaps, 1SG nonpast verbs in Russian.[6] Representative examples are given again in (9):

(9) *muč/*muščju 'I stir up'
     *očučus'/*očuščus' 'I find myself'
     *pobežu/*pobeždu 'I win'
     *erunžu/*erunždu 'I behave foolishly'
     *lažu/*lažd'u 'I climb'

In addition to the two broad classes of proposals pioneered by Halle (1973), it has sometimes been suggested that ineffability here is driven by homophony avoidance. For instance, one would expect both *lazit'* 'to climb' and *ladit'* 'to be on good terms with' to have the same 1SG nonpast form, e.g., *lažu*. Yet as Halle (1973) already observes, there are other verbs in the language for which homophony avoidance does not result in gaps (cf. *vožu* 'I lead' or 'I cart'); see also Hetzron 1975 and Sims 2006. Nor can the missing inflections be attributed to phonetic or phonological factors, as suggested in the theoretical treatment of gaps (e.g., Fanselow and Féry 2002; Orgun and Sprouse 1999). Russian has

---

6 I thank Jennifer Preys for her assistance in the Russian study. Her research was supported by a Penn Undergraduate Research Mentoring (PURM) fellowship.

5.1 Finding Gaps

1sg. forms such as *vonžu* 'I thrust (a knife)' or *šuču* 'I joke': well-formedness conditions do not appear to be the plausible explanation (Sims 2006).

Russian verbs have two primary conjugation classes, which are defined in terms of the third-person plural morpheme and the theme vowel: /e/ and /o/ for the first conjugation and /i/ for the second. Verbs in the second conjugation have a palatalization alternation targeting verbs whose stem ends in a dental consonant; interestingly, all 1SG nonpast gaps appear in the second conjugation and all contain a stem-final dental. My analysis thus targets those verbs with a stem-final *t*, which can be palatalized as either *č* (e.g., *metit'-meču* 'mark') or *šč* (e.g., *smutit'-smuščú* 'confuse') in the 1SG nonpast. The Zaliznjak (1977) morphological dictionary was used to count the frequencies of the two mutations of /t/ as *č* and *šč*. Verbs that share a stem with other forms but differ in the presence or absence of a prefix or the reflexive suffix -*sja* are counted only once. This is because Russian verb stems have multiple variants that differ in prefixation; Pesetsky (1977) and Baronian (2005) observe that variants of the same stem either share the same mutation in the 1SG nonpast — *č* or *šč* — or are both ineffable. As in the other case studies, verb roots that occur less than once per million tokens in the Russian National Corpus (Sharoff 2005) were excluded. This yields fifty-nine attested second-conjugation stem-final *t* roots. In the 1SG nonpast, thirty-seven of them follow the majority pattern, mutating to *č*, twenty roots instead mutate to *šč*, and two remaining roots follow other idiosyncratic patterns. While the *č* mutation is clearly the majority pattern, it fails to reach the productivity threshold, which can tolerate no more than $\theta_{59} = 14$ exceptions. If we include all words in the corpus regardless of their frequency, the total number of verb stems increases to sixty-six, with the same twenty-two non-*č*-alternating exceptions, still too many to support any productive process because $\theta_{66} = 16$.

Thanks to the diffused distribution of stem alternations, no single process survives the Tolerance test. Learners must lexically mark the specific palatalization process of each verb attested in 1SG nonpast; the verbs without attested inflectional forms fall into gaps because there is no regular or default process to pick them up. The classic problem of Russian gaps is therefore resolved, again on a purely numerical basis.

5.1.4 The Indeterminacy of Polish Masculine Genitives

The case of Russian is notable because, unlike *stridden* and the gaps in Spanish verbs, which are few and sporadic, a non-trivial number of verbs show inflectional defectiveness. The case of Polish masculine genitives, however, must take the crown for ineffability, even though it has not featured in the defective gap literature.[7]

The Polish masculine genitive system has been put forward in the language acquisition literature as a challenge to the dual-route model of morphology, according to which learners must seek, and find, a default rule. In Polish, masculine nouns in the genitive singular either take the *-a* or *-u* suffix, but Dąbrowska (2001, 560) shows convincingly that neither is the default when assessed with the typical tests for productivity (e.g., Marcus et al. 1995). Of the two, one is necessarily the statistical majority (*-a*), but it fails to function as a default. In contrast to the past-tense *-d* in English, which is always used for novel verbs, some novel and low-frequency masculine singulars in Polish add *-a* for the genitive but others add *-u*. Loanwords also take on *-a* and *-u* in an unpredictable fashion. In addition, phonological conditioning of the suffix choice appears absent (Mausch 2003). The distribution of these two suffixes has been extensively studied but has consistently defied a systematic classification. The choice of the suffix appears arbitrary ("really depends on usus [sic]"; Westfal 1956, XV), and speakers are often unsure about which form to use. While some masculine nouns show *-a* and *-u* in free variation (e.g., *deseni* 'design'), for others, the genitive singular is ineffable: "A Pole is often uncertain as to the correct genitive form of cities like *Dublin* 'Dublin' and *Göteborg* 'Gothenburg'. [...] My informants were uncertain as to the correct genitive form of *Tarnobrzeg* 'Tarnobrzeg' which, according to the dictionaries, takes *-a*" (Kottum 1981, 182f.). For instance, the native speakers of Polish I consulted are reluctant to use either *-a* or *-u* for the genitive singular form of the following masculine nouns, despite the reported tendency that animate nouns prefer *-a* and abstract nouns prefer *-u* (Dąbrowska and Szczerbinski 2006; Westfal 1956):

(10) drut 'wire'
 rower 'bike'
 balon 'balloon'

---

[7] I thank Margaret Borowczyk for her assistance in this study.

5.1 Finding Gaps

karabin 'rifle'
autobus 'bus'
lotos 'lotus flower'

These distributional facts collectively point to the absence of a productive suffix; which suffix a noun selects must be learned by fiat, a process that takes the learner well into the teenage years (Dąbrowska 2001, 2005). In contrast, the genitive plural for the masculine nouns is more conventional: the default suffix is *-ów* with a small number of exceptional nouns taking *-i/-y*.

The application of the Tolerance Principle is again very straightforward. For the masculine singular genitive, it must be the case that neither *-a* nor *-u* reaches the requisite threshold to tolerate the other as exceptions. For the plural, by contrast, it must be the case that the number of *-ów* suffixed items thoroughly overwhelms the number of *-i/-y* taking items. In an analysis of a corpus of thirty hours of child-directed Polish, Dąbrowska and Szczerbinski (2006) find forty *-a* and eighteen *-u* suffixed genitives: clearly, neither meets the threshold of productivity because the maximum number of exceptions is only $\theta_{58} = 14$. To expand the lexical database, I analyzed the nouns from child-directed Polish (available in CHILDES). A native speaker of Polish, in consultation with dictionaries and other speakers, manually classified masculine nouns into four classes according to their genitive case endings (SG: *-a* or *-u*; PL: *-ów* or *-i/-y*).[8] The results are shown in Table 5.4, along with children's error rates reported in Dabrowska's 2001 study.

Several interesting patterns emerge from the combined analysis of the input and children's case marking. The Tolerance Principle correctly identifies the absence of a productive suffix in the singular and the presence of a productive suffix in the plural. In the singular, while *-a* is clearly the majority (837 of 1,353), the alternative suffix *-u* (516) is far too numerous to be tolerated as exceptions for *-a* to be the productive default: the critical threshold is only at $\theta_{1353} = 188$. Consequently, speakers must encounter attested instances of suffixation in order to learn the correct form of case marking; failing to do so results in the indeterminacy noted by previous scholars and in the native speaker data I collected (10). In the plural, by contrast, the *-ów* suffix has no

---

8 The tabulation is complicated slightly by a small number of masculine nouns that take *-a* in the GEN.SG and *-u* in the DAT.SG. Since the word list does not provide a way to determine the case of individual tokens, the count of *-u* nouns (516) is likely be a slight overestimate, but it is unlikely this would change the results, because the number of nouns potentially affected is rather small and both classes are quite far from the productivity threshold.

Table 5.4

Distributions of genitive suffixes on Polish masculine nouns, the productivity predictions of the Tolerance Principle, average frequency (mean number of tokens per million words), and children's error rates

	Suffix	Types	Productive?	Avg. freq.	Child error rate
GEN.SG:	-a	837 (62%)	No	7.2	1.28%
	-u	516 (38%)	No	8.8	0.24%
GEN.PL:	-ów	551 (90%)	Yes	6.5	0.41%
	-y/-i	61 (10%)	No	11.4	15.53%

difficulty withstanding sixty-one exceptions, which is below the threshold of $\theta_{612} = 95$.

These numerical predictions are strongly supported by children's acquisition of the genitive. For the singulars, the lack of a productive process offers no opportunity for overregularization and children's error rates on both suffixes are very low. For the plural, however, the existence of the -ów productive default serves as the attractor for overregularization: to wit, the irregular -y/-i suffixed items have by far the highest error rates due to the productive application of the -ów suffix, even though they have a higher average *token* frequency in child-directed Polish. This asymmetry in the acquisition of productive and unproductive processes is exactly what the crosslinguistic study of morphological acquisition reveals (section 2.3.3). The presence of a productive rule for generating the genitive plural leads to overregularization errors, whereas the lack of such a rule in the genitive singular leads to defectivity.

The four case studies of ineffability have been given a unified analysis by the Tolerance Principle, which correctly predicts where gaps should emerge. My account formalizes the insight from previous proposals that indeterminacies result from competing alternations. Furthermore, the present theory of gaps has strong continuity with the study of productive morphological processes (chapter 4), which suggests that a fundamental dichotomy between gaps and rules needn't be postulated in the architecture of the grammar but emerges from the input data and the generalization process in language acquisition.

I now turn to the dynamic aspects of productivity in language variation and change.

5.2 The Rise and Fall of Productivity

Proposed as a solution for the productivity problem, the Tolerance Principle must be a component of the human cognitive system that governs all linguistic matters regardless of time or place. One approach to validation, which I have pursued in the preceding pages, is to examine a wide range of crosslinguistic phenomena and the processes by which children acquire them. In those cases, the target state of the linguistic systems is fairly well understood, which forms a stable point of comparison against the predictions of the Tolerance Principle. Yet another, and possibly more interesting, application can be found in historical linguistics. I contend that the Tolerance Principle is a causal force that shapes the history of languages. As such, it has the potential of uncovering the deterministic factors in language change.

Once again, I use morphological problems to illustrate the application of the Tolerance Principle in language variation and change, but the method is general and can be applied to both phonology and syntax. By morphological change, I refer to changes in the morphological membership of lexical items. Traditionally, two types of morphological changes have received special attention (e.g., Campbell 2004):

(11) a. Analogical leveling, where a word shifts from a more restrictive class to a less restrictive one: the past tense of *row* used to be *rew*, along with *blow-blew*, *grow-grew*, etc., and is now *rowed* (regular)

   b. Analogical extension, where a word shifts from a less restrictive class to a more restrictive one: the past tense of *wear* used to be regular (*werede*) but took on the form of *wore*, following the class of irregular verbs such as *bear-bore* and *swear-swore*

In my view, both kinds of changes result from the application of rules to words that previously did not fall under their reign but nevertheless could have. That is, the rule/class on which leveling and extension operate must be productive. We write R^+ to indicate that R is productive; an unproductive rule (R^-) is one that applies to a fixed set of words and nothing else, and its membership must be somehow memorized during the course of language acquisition. This amounts to claiming that, for instance, when *werede* became *wore*, the *bear-bore* pattern (strong verb class IV) must have been productive such that it could assimilate words that fit its structural description. The strongest position one can take here is to deny lexical analogy in the absence of productivity, at least as an endogenous means of change.[9] I am tempted to take such a position because lexical/nonproductive patterns are almost never overgeneralized (section 2.3) by children during the normal course of language acquisition. Of course, when learners experience deviates from the normal course, the productivity of rules may change, and language change follows.

The prima facie evidence against the productivity-based approach to change would be cases where an item has shifted from an undoubtedly productive rule to an undoubtedly nonproductive rule/pattern. The most detailed study of this type is a quantitative analysis by Anderwald 2013, who examined the usage trajectories of *thrive-throve*, *dive-dove*, *plead-pled*, *drag-drug*, and *sneak-snuck* in the Corpus of Historical American English (COHA; Davies 2010). Of these, *throve* is a clear case of leveling, gradually falling into disuse and replaced by *thrived*, the regular and productive form for the past few centuries. The other four verbs are apparently instances of extension — which should not be possible since the irregular rules have not been productive in at least two hundred years. But three of the four counterexamples are only apparent. Two of these verbs (*dove* and *pled*) had variable strong and weak forms from the beginning of the corpus; the morphological changes in these cases have been a matter of frequency fluctuation, which is well attested in the history of English (Taylor 1994). The frequency of *drug* has always been very low and never surpassed 1% during any decade of the COHA Corpus, with *dragged* the overwhelmingly dominant alternative.

Only *sneak-snuck* seems to be a genuine case of extension along an irregular pattern.[10] It was virtually unattested in the corpus prior to 1900 and is currently in a dead heat with the regular form *sneaked* (Anderwald 2013, Figure 4). But the history of *snuck* is more complicated, and it does not clearly constitute a counterexample to the hypothesis that only productive rules can assimilate new members. The emergence of *snuck* is "of doubtful origin" according to the OED. A verb *snícan*, most likely a strong verb, had been present in Old English, which raises the possibility that *snuck*, or its ancestral form, had been present all along rather than being a genuine instance of innovation along an unproductive pattern. However, there is a big gap of about 500 years from around 1100 until 1598 (in Shakespeare) in which no forms at all are attested, followed by another three centuries in which only spotty occurrences

---

9 No one can predict what external forces — prestige, power, punishment — may influence the course of language change and how they may do it.

10 I thank Don Ringe for enlightening discussions about *snuck*; the usual disclaimers do not apply.

5.2 The Rise and Fall of Productivity

of *sneaked* can be found, until late nineteenth century U.S. English, as shown in Anderwald's COHA analysis.

However, in addition to viewing *snuck* as a genuine case of irregularization, there are still several possibilities to consider. The very first occurrence of *snuck*, evidently one that Anderwald missed, dates back to 1889 in Eugene Field's *A Little Book of Western Verse*. In a (disturbing) piece titled *A Proper Trewe Idyll of Camelot*, we find:

Then, looking down beside him, lo! his lady was not there—

He called, he searched, but, Goddis wounds! he found her nonywhere;

And whiles he searched, Sir Maligraunce rashed in, wood wroth, and cried,

"Methinketh that ye straunger knyght hath snuck away my bride!"

Perhaps *snuck* had been around despite not having been recorded in historical corpora: there is no irregular innovation. Whatever literary license Field had taken, perhaps even a racist or misogynist kind (Don Ringe; personal communication), he might have inadvertently help spread the form, *snuck*, because his writings were evidently very popular in his days. If so, then the rise of *snuck* would be more social rather than linguistic and thus needn't concern a formal theory of language learning and change. In light of the uncertainties surrounding the origin of *snuck*, I maintain, as a working assumption at least, that productivity is the sole driving force in morphological change, echoing earlier proposals in the generative approach to morphological change (e.g., Anderson 1988a; Kiparsky 1974 and especially Anderson 2015 which also stresses the importance of learning in language change). As before, I assume that the Elsewhere Condition governs the application of rules to lexical items. Should a more specific and applicable rule become productive, it has the potential of assimilating additional members that fall under its structural description.

Following this line of reasoning, morphological change must be driven by the numerical relationship between N and e for any given rule R: R^+ if $e \leq \theta_N$ and R^- if $e > \theta_N$. This process is depicted in Figure 5.2.

Many factors may change the values of N and e for a rule. I have already discussed the effect of sampling in language acquisition in chapter 4—that a rule may be productive for some language learners but unproductive for others. One may construct a formal model of rule change as a dynamical system of $(N, e)_t$ that undergoes perturbation over generations (t); it is quite clear that the system will bifurcate as a threshold function of N and e (see Yang 2006b

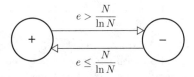

Figure 5.2
The change of rule productivity as a function of N and e

for details). Once a rule becomes productive (R^+), the exceptions to the rule can only be maintained by fiat on a lexicalized list, thereby opening up the possibility of leveling to R^+. If, by contrast, a productive rule becomes unproductive (R^-) as the number of exceptions crosses the Tolerance threshold, then all its members will be subject to leveling to a yet more general (and productive) rule, leading to the collapse of a structural class. The recent changes in the Icelandic morphosyntactic system are a case in point.

5.3 Diagnosing Sickness

I investigate the morphosyntactic change in Icelandic known as Dative Sickness.[11] Following an overview of the empirical phenomenon, we show that with the use of both historical and modern corpus statistics, the Tolerance Principle provides a mechanistic account of how the change was actuated and why it took place in the observed direction.

5.3.1 The Symptoms

Icelandic has witnessed a recent change, prescriptively referred to as Dative Sickness and more neutrally as Dative Substitution, which is quite familiar across the Germanic languages (Smith 1994). Experiencer subjects[12] in certain verbs and predicates that traditionally bear the accusative case have now changed, or are changing, to the dative case. (12) shows some representative examples (Jónsson 2003).

---

11 This work would have been possible without the help from Iris Edda Nowenstein and Einar Freyr Sigurðsson, for which I am grateful.

12 These are subjects that receive sensory or emotional experience with respect to the action or state described by the verb (e.g., *hurt*) or predicate (e.g., *happy*).

5.3 Diagnosing Sickness

(12) a. **Mig**   langar að fara heim
       me-ACC wants to go home
      Mér langar að fara heim
      me-DAT wants to go home
      'I want to go home.'

    b. **Þá**   vantar fleiri stóla
       them-ACC needs more chairs
      Þeim vantar fleiri stóla
      them-DAT needs more chairs
      'They need more chairs.'

    c. **Stelpuna** kitlaði í tána
      the girl-ACC tickled in the toe
      Stelpunni kitlaði í tána
      the girl-DAT tickled in the toe
      'The girl was ticklish in the toe.'

Dative Substitution is a recent change that started in the nineteenth century. The well-documented history of the Icelandic language, from which the relevant lexical statistics can be gathered, provides a solid testing ground for the Tolerance Principle as a model of language change. Traditionally, accusative subject marking appears to have had some productivity for experiencer verbs which, as a class, have shared lexicosemantic properties. For instance, some experiencer verbs first attested in the Old Icelandic period (prior to 1350) took on accusative subjects; these include *hrylla við* 'be horrified by', *óra fyrir* 'dream of' and *rámá* 'have a vague recollection of' (Jónsson and Eythórsson 2011, 224):

(13) a. **Nemendurana**   hryllir   við þessari tilhugsun
       the.students-ACC horrifies at   this   thought
       'The students are horrified by the thought of this.'

    b. **Engan**   hefði getað órað   fyrir þessu
      nobody-ACC had   could dreamed for   this
      'Nobody could have dreamed of this.'

    c. **Mig**   rámar   í að hafa hitt hann einu sinni
      me-ACC recollects in to have met him   one   time
      'I have a vague recollection of having met him once.'

And at least one loan verb *ske* 'happen', for which the oldest example dates back to the late fourteenth century, also took accusative subjects. The following example is from mid-seventeenth century:

(14) eins og **mig** hafði skeð fyrir átta árum
as me-ACC had happened for eight years
'As had happened to me eight years earlier'

Finally, there are verbs that originally took nominative subjects (*vona* 'hope', *skynja* 'sense') and dative subjects (*klæja* 'itch') in Old Icelandic but later became associated with accusative case.

(15) Nominative subject in Old Icelandic but accusative subject now

a. Þá vonar **mig** að þær smámsaman fjölgi
then hopes me-ACC that they gradually increase
'Then I hope that they increase in number.'

b. **mig** skiniar ecki sannara en seigi
me-ACC senses not true than say
'I do not sense more truthfully than I say.'

(16) Dative subject in Old Icelandic but accusative subject now

a. því **mér** klæjar þar mjög
since me-DAT itches there much

b. því **mig** klæjar þar mjög
since me-ACC itches there much
'because I am itching there so much'

The important point about these examples is that accusative subject marking appears productive at an earlier stage of Icelandic; as such, it was able to attract experiencer verbs that previously had nonaccusative subjects. This is no longer possible. Instead, accusative subjects are now losers and have been subject to substitutions by other case forms. In addition to Dative Substitution, some of the verbs and predicates that traditionally took accusative subjects may level to the nominative, as can be seen in a survey of native speakers (Table 5.5).

At some point in the history of the language, accusative case marking ceased to be productive; we would like to understand how and why.

Verbs with accusative theme subjects (e.g., *reka* 'drift'), however, tell a very different story. First, this class has never shown much productivity; in fact, they have been steadily falling into disuse. Jónsson and Eythórsson (2011) list

5.3 Diagnosing Sickness

Table 5.5
Verbs originally taking accusative experiencer subjects in Icelandic; total results in % of case marking. (From Jónsson and Eythórsson (2005, 232))

Verb	Gloss	NOM	ACC	DAT	Other
gruna	'suspect'	7.0	65.4	27.2	0.3
dreyma	'dream'	9.3	64.7	25.4	0.5
langa	'want'	1.4	58.5	39.8	0.3
minna	'seem to remember'	21.8	53.0	24.9	0.3
vanta	'lack, need'	1.8	52.2	45.4	0.6
svíða	'smart, sting'	3.2	43.4	52.9	0.5
svima	'feel dizzy'	3.0	36.4	60.4	0.2

seventy-seven accusative theme subject verbs in Old Icelandic, a list that has been gradually shrinking, with only ten to fifteen still regularly used in Modern Icelandic. Second, and more interestingly, the accusative theme subjects have been the victim of *Nominative* Substitution rather than Dative Substitution (Eythórsson 2002; Jónsson 2003):

(17) a. **Bátana** rak að landi
the.boats-ACC drifted to land

b. **Bátarnir** ráku að landi
the.boats-NOM drifted to land

This is significant because there in fact *are* verbs that take dative theme subjects (18a). Yet they not only fail to attract the accusative subjects but also fall under nominative substitution themselves (18b):

(18) a. **Leikjunum** lyktaði með jafntefli
the.matches-DAT ended with draw

b. **Leikirnir** lyktuðu með jafntefli
the.matches-NOM ended with draw
'The matches ended in a draw.'

Taken together, we can summarize the current state of variation and change in subject case marking (Jónsson 2003, 152):

(19) a. Theme subject: accusative/dative → nominative (Nominative Substitution)

b. Experiencer subject: accusative → dative (Dative Substitution)

Theoretically, both Nominative Substitution and Dative Substitution can be regarded as instances of leveling: a lexically specific and more marked form drifting toward a more general and less marked form. I will not review the extensive arguments that the nominative case is the unmarked case in Icelandic (see, among others, Schütze 2001; Zaenen, Maling, and Thráinsson 1985). If so, then the nominative as a destination of leveling is sensible. Furthermore, Jónsson and Eythórsson (2005, 232) report a frequency effect in Dative Substitution: the less frequent verbs in Table 5.5 (e.g., *svíða* 'sting', *svima* 'feel dizzy') have the highest dative usage. Finally, studies of language acquisition in Icelandic (Nowenstein 2015; Sigurjónsdóttir 2002) show that children initially acquire, and overgeneralize, the nominative case for subjects. The dative case then follows, which is also overgeneralized for experiencer subjects. Accusative subjects, which are lexically idiosyncratic, are acquired last. This sequence of case marking acquisition is most likely due to the frequency effects in the input. The nominative case is by far the most dominant, both statistically and structurally (see especially Schütze 1997 for discussion). And there is a relatively small number of accusative subject verbs—I'll come back to the numbers momentarily—that are already in variation with dative subjects (see Table 5.5), which contributes to their late acquisition.

Overall, then, the case marking system in Icelandic can be viewed as a hierarchy of forms governed by the Elsewhere Condition (e.g., Legate 2008; Marantz 2000, Yip, Maling, and Jackendoff 1987). The default structural case is nominative. For the experiencer verbs, the default (inherent) case is dative but for an exceptional and lexicalized list, the accusative case is used. When the accusative marking for a particular verb is lost (i.e., imperfectly acquired), the next option is the dative, which is a more specific match than the most general nominative form. For the theme subjects, however, it appears that neither the accusative nor the dative is capable of absorbing the other; when they level, it is straight to the most general nominative form. But why should experiencer and theme subjects head toward different destinations? We must also confront the critical question in the study of change, the Actuation Problem (Weinreich, Labov, and Herzog 1968): Why did the subject case leveling take place in the nineteenth century, not before or after?

5.3.2 Predicting Case Substitution

The Tolerance Principle provides a very simple account for the different trajectories of change for theme subjects and experiencer subjects. The answer is again in the numbers. Table 5.6, which is adapted from Eythórsson 2002, provides the distribution of subject case in Modern Icelandic according to thematic roles.

Table 5.6
The frequency of accusative and dative subjects in Modern Icelandic

	ACC	DAT	Total (N)	θ_N
Theme	14	19	33	9
Experiencer	37	227	264	47

This table alone explains the difference between the leveling of theme and experiencer subjects. For theme subjects, neither accusative nor dative marking can become the productive inherent case since the maximum number of exceptions is only $\theta_{33} = 9$. Thus, learners must lexicalize, for each specific verb or predicate, whether the subject takes the accusative or dative case. When a verb fails to get on the list—lexicalization takes repeated exposure and is prone to error—it can only level to the next more general case form, which is the nominative, the most general structural case in the language. This accounts for Nominative Substitution (19a).

For experiencer subjects, however, the dative is the productive case because it can easily tolerate the thirty-seven accusative subjects: the threshold value is $\theta_{264} = 47$. This accords well with the observation by Eythórsson (2002, 207) that "within the limited class of verbs taking oblique subjects, experiencer verbs with dative subjects far outnumber those with accusative subjects in standard Icelandic." Thus, if a lexically marked accusative is to level, it will fall under the reign of the dative, which is a more specific match than the nominative and is thus favored by the Elsewhere Condition. This accounts for Dative Substitution (19b).

The story isn't quite complete. We need to address the apparent counterexamples of accusative experiencers taking on the nominative form as shown in Table 5.5. There is some controversy in the literature on this matter. Jónsson (1997) does not consider this possible while the survey results in Table 5.5 as well as earlier work (e.g., Halldórsson 1982) clearly show a non-negligible rate of nominative marking for experiencer subjects.

In my view, a likely account of this state of affairs is that the nominative forms are historical residues. Recall that Dative Substitution was a nineteenth century phenomenon. Prior to that point, accusative experiencers showed variation with the nominative, dating back to Old Icelandic. The verb *langa* 'want', for instance, generally occurs with the accusative but appeared with nominative subjects as far back as *The Sagas of Icelanders*:

(20)   Orkneyíngar   myndi   lítt lánga til, at hann kæmi
       Orkneymen-NOM would-3PL little want to that he   came
       vestr   þagat
       westward thither
       'The men of the Orkneys would not be eager for him to come here to the west.' (*Fornmannasögur* VII.28)

As Eythórsson (2002, 203) notes, by Early Modern Icelandic (sixteenth and seventeenth centuries, well before the emergence of Dative Substitution), many accusative subjects were attested in the nominative, including *dreyma* 'dream', *gruna* 'suspect', *hungra* 'feel hungry', *kala* 'suffer frostbite', *langa* 'want', *skorta* 'lack', *undra* 'wonder', *vanta* 'lack, need', *verkja* 'hurt', *þystra* 'feel thirsty', and so on. And the accusative-nominative variation is also well attested in the textual records of Modern Icelandic (of the nineteenth and twentieth centuries). Since these forms were/are still used, the learner would naturally retain a level of usage. Furthermore, recall that Icelandic-learning children acquire the nominative case first as the structural default. Yet it is still notable that once Dative Substitution became a possibility thanks to the emergence of the dative as a productive form of case marking, it very quickly dominated the nominative option and is now the majority pattern for the lower-frequency experiencer verbs (Table 5.5). The claim of the dative as the default for experiencer verbs could be further tested. For instance, novel experiencer verbs and predicates would be expected to take the dative rather than nominative (or accusative) subject — a prediction awaiting further investigation.

5.3.3   The Actuation of Change

All of this sets up the critical question: How and why did the dative become the productive form for experiencer subjects? The situation for experiencer subjects prior to the emergence of the dative default seems comparable to that for theme subjects in Modern Icelandic: while the accusative and dative coexisted, neither was able to emerge as the default so both leveled to the nominative.

5.3 Diagnosing Sickness

According to the Tolerance Principle, a rule with its associated values (N, e) can gain productivity only if the numerical relation between N and e changes. If N gets larger, the rule may tolerate the exceptions previously deemed too numerous. Alternatively, e becoming smaller may also push an unproductive rule into the region of productivity. In either case, one would require accurate lexical statistics that reflect the native speaker's grammatical knowledge at or just prior to the time of change, which is nigh impossible given the limitations of historical corpus data.

But several indirect means of investigation suggest that the emergence of Dative Substitution was most likely due to a general vocabulary reduction of experiencer verbs and predicates, which appeared to have affected the accusative subjects more than the dative subjects.

The first line of argument concerns the individual learner's effective vocabulary and its effect on the productivity of case marking. Many of the experiencer verbs listed in previous studies (e.g., the statistics reported in Table 5.6) are already very rare in spoken Icelandic and very unlikely to be completely acquired by native speakers. I conducted an informal survey to see how much of the traditional experiencer vocabulary is still retained by Modern Icelandic speakers. Iris Edda Nowenstein compiled a set of ninety-three traditionally dative and ninety-nine traditionally accusative experiencer verbs listed in Jónsson 2003, in addition to ninety-six dative subject predicates (e.g., *Mig hunger* 'me-DAT hunger' or 'I am hungry') from *Íslensk tunga III* (Thráinsson 2005). We then asked several native speakers to identify those they recognize on the checklist. The results are summarized in Table 5.7.

Table 5.7
Vocabulary estimates of prescriptive experiencer verbs and predicates in Modern Icelandic

	DAT subjects	ACC subjects	N	θ_N
Speaker 1	106	**28**	134	**27**
Speaker 2	129	**38**	167	**32**
Speaker 3	108	**32**	140	**28**
Speaker 4	60	**24**	84	**18**
Speaker 5	56	**16**	72	**17**
Speaker 6	63	**18**	81	**18**

Speakers 1, 2, 3, and 4 received their college education in Iceland and were, at the time of writing, doctoral candidates in linguistics. Speaker 5 acquired Icelandic as a young child before moving to the United States but retained use of the language in family situations including frequent trips to Iceland. Speaker 6 is a six-year-old child, and the estimates are provided by her stepmother (Speaker 3). As Table 5.7 shows, all speakers, especially the younger ones, are right around the cusp of productivity predicted by the Tolerance Principle (see the values in columns 3 and 5), supporting the conclusion that the dative is the default case for experiencer subjects with a tolerable number of accusative subject exceptions. It is clear that the experiencer verbs and predicates have significantly fallen into disuse because none of our speakers approaches the dictionary-based figures in Table 5.6.[13] Conceivably, this is how Dative Substitution emerged in the history of Icelandic. Initially, there are relatively large numbers of accusative (e) and dative ($N - e$) experiencer verbs and predicates such that neither form could emerge as the default. However, over time the values of e and N were gradually reduced—perhaps at different rates; see below—such that by the nineteenth century the numerical relationship between N and e started to support the productivity of dative subject marking.

The second line of argument comes from lexical statistics in historical corpora. To formally establish this possibility requires more historical work that can be undertaken here but we are in a position to offer a preliminary assessment. Einar Freyr Sigurðsson helped me to extract all the accusative and dative subject experiencers from the Icelandic Parsed Historical Corpus (IcePaHC; Wallenberg et al. 2011) in two periods immediately prior to the emergence of Dative Substitution. From 1725 to 1791 and a text of 86,000 words, 22 accusative and 60 dative experiencer verbs are found with three showing variable forms in both sets. From 1830 to 1882 and a text of 104,000 words, the numbers are 22 and 62 respectively, with one appearing in both sets. (The IcePaHC does not contain data from 1792 to 1829.) Though these numbers ($N = 80/82, e = 22, \theta_N = 18$) already approach the threshold for productivity, many of the verbs are attested too sparsely (e.g., only once) to inspire confidence in the statistical conclusions.

---

13 Except for Speaker 2, who recognizes more accusative experiencer verbs (38) than those listed in Modern Icelandic dictionaries (37; Table 5.6). Not exactly a surprise that the speaker is currently writing a dissertation on Icelandic argument structure.

5.3 Diagnosing Sickness

Rather more suggestive, however, are the *token* frequencies of these verbs. On average, the dative experiencer verbs are almost twice as frequent as the accusative experiencer verbs for both periods. On the assumption that more frequent words tend to be acquired more reliably, we conjecture that while both types of subjects were falling into disuse, the less frequent accusative subject verbs were more significantly affected — that is, they are less likely to be acquired by language learners over time. This is consistent with the informal survey results in Table 5.7: although the checklist contains roughly equal numbers of traditionally accusative and dative experiencer verbs (99 vs. 93), a significantly greater proportion of the former were lost than the latter (74% vs. 55%, averaged across the six survey subjects).

To summarize, the Tolerance Principle, which is independently motivated by a large number of studies presented so far, makes correct predictions about the distribution, directionality, and time course of case substitutions in the history of Icelandic. The numerical basis of its application provides a concrete account of the driving forces of change, and it can help sharpen research questions to be explored in more targeted fashion in future research. It is worth emphasizing again that the current program of language change (Figure 5.2) is general and not limited to morphological change. The rises and falls of grammars have long been recognized as the cause of language change (Paul 1888), with language acquisition brought to the fore in the generative tradition initiated by Halle 1962 and Lightfoot 1979. According to this view, the child selects the simplest grammar compatible with the input data according to the Evaluation Metric (Chomsky 1955, 1965; Chomsky and Halle 1968), a process that has been described as "reanalysis" in the traditional literature. Yet the notion of simplicity in the context of language acquisition has never been successfully developed. In the extreme, a maximally simple grammar may encounter numerous exceptions while a maximally complex grammar may be descriptively perfect by making an exhaustive listing of everything — obviously neither position is tenable. I propose that the Tolerance Principle offers a metric by which a grammar may be deemed "good enough". The child proceeds by evaluating grammars with increasing complexity, stopping at the grammar that passes the Tolerance Principle in the process that figures prominently in the acquisition studies (chapter 4, especially (2)). The Icelandic study presented here provides a template for an integrated approach to language acquisition and change.

**

From the perspective of the Tolerance Principle, the case studies in this chapter fall along the same continuum with the "normal" state of language where the supply of novel expressions has no limit. Perturbations in vocabulary composition, which may be due to external forces, can lead to profound changes in the grammar. An elucidation of the synchronic process of language learning, which is now studied with increasing sophistication and precision, will undoubtedly help to isolate the causal factors in language variation and change (Kiparsky 1965; Labov 1989; Lightfoot 1979).

This chapter has been a catalog of accidents and contingencies: gaps, sicknesses, and other linguistic failings are not a matter of necessity. The fact that they exist alongside productive processes in language has led to considerable discomfort for theories that attempt to encompass all linguistic problems within Universal Grammar. A novel contribution of the Tolerance Principle is to create a new division of labor between Universal Grammar and experience, a theme that I develop further in chapter 6.

6 The Logic of Evidence

Imagine being shipwrecked on a desert island. If you come across ten exotic species, seven of which are tame and friendly, you'll probably assume the next encounter is harmless: seven out of ten seem pretty good odds. But one out of ten? Two out of ten? You'd be well advised to proceed with caution.

Generalizations evidently require the weight of evidence: seven exemplars appear sufficient but two probably won't do. This is not to say that the sufficiency of evidence is foolproof: the other three species, with which we have had no direct experience, may well turn out to be dangerous. Nor does sufficiency guarantee permanence: if an additional ten species have come our way, suddenly seven out of twenty no longer inspire confidence.

So far so good. Similar kinds of learning must be going on all the time, where generalizations extend from attested examples to an entire class. Do dinosaurs have wings? Will leaves turn yellow in the fall? Must all Jedis wield a lightsaber? In this chapter, I argue that children follow a similar logic of inference, which will dissolve some of the hardest cases in the study of language. But first, let's see why generalization in language learning is uniquely interesting and challenging.

****

Linguistics took a psychological turn thanks to a logical conundrum. Chomsky's critique of Skinner's behaviorist program (1957) brought the indeterminacy of induction (Goodman 1955; Quine 1960) to the center stage of language and cognition. The associationist approach to language is fatally undermined by the indeterminacy of stimulus and response. A "Dutch" painting may elicit "'Clashes with the wallpaper', 'I thought you liked abstract work', 'Never saw it before', 'Tilted', 'Hanging too low', 'Beautiful', 'Hideous', 'Remember our camping trip last summer?', or whatever else might come into our minds when looking at a picture. ... We cannot predict verbal behavior in terms of the stimuli in the speaker's environment, since we do not know what the current stimuli are until he responds" (Chomsky 1959, 31). As Lila Gleitman once memorably put it, a picture is worth a thousand words *and that's the problem.*

The indeterminacy of induction is particularly acute in the acquisition of syntax. The grammar must be projected from a finite amount of data; logically, however, infinitely many hypotheses can be formed. Worse still, the absence of negative evidence sets language acquisition apart from many other learning problems, where the hypotheses can be rejected or revised when the learner receives useful feedback. To successfully learn a grammar, then, children must get help from within as well as without. They may come equipped with a very constrained space of hypotheses (Universal Grammar), which considerably sidesteps the problem of induction. At the same time, it is also possible that the learning mechanisms for language have some clever ways of squeezing a bit more out of the limited positive data. Indirect negative evidence (Chomsky 1981) is a powerful idea that researchers have come back to over and over.

In this chapter, I first discuss the perils of indirect negative evidence as it is typically invoked in language acquisition research. As an alternative, I introduce the notion of *sufficient positive evidence*, a decision rule on the validity of generalizations. As on the desert island, learners encounter a well-defined set of N items of which a subset of M items are explicitly attested with some property P. Should P be generalized to the remaining $(N-M)$ items and beyond, to novel items that are similar to N? Or should P be restricted to those for which there is explicit positive evidence? I suggest that the decision needn't take place right away, before a sufficient amount of evidence exists one way or the other. The quantity of sufficiency, as readers might have guessed, is provided by the Tolerance Principle again.

The execution of sufficient positive evidence will be illustrated with two empirical cases that traditionally have been treated with indirect negative evidence and its surrogates — but unsuccessfully as we will see. The end result is not only a solution to some persistent puzzles in language acquisition but also a considerable reduction of the innate machinery of cognition that has been assumed to be the prerequisite in the traditional literature.

6.1 Inference and Weight of Evidence

To learn a grammar, generalization is a matter of necessity. Language is infinite but language experience is finite and sparse (chapter 2). That children consistently go beyond the input is made evident even, and especially, when they make mistakes. Every time a child says *Don't giggle him* or *I said her no*, there are grammatical generalizations at work, even when they occasionally get it wrong.

6.1.1 Indeterminacy

How to generalize "just right," as Goldilocks would say, has been the central concern for learnability research. Constraining the hypothesis space is an effective approach, one taken in theoretical linguistics such as the principles-and-parameters framework (Chomsky 1981) and Optimality Theory (Prince

6.1 Inference and Weight of Evidence

and Smolensky 2004) as well as in formal studies of computational learning (Valiant 1984; Vapnik 2000). Learners can never go out of bounds, which greatly reduces the problem of indeterminacy. The scope and limits of grammars are often best viewed from the perspective of child language acquisition. On the one hand, there are logically possible hypotheses about language that learners never seem to entertain (Crain 1991; Legate and Yang 2002; Smith and Tsimpli 1995; Tettamanti et al. 2004). On the other, children spontaneously create expressions that are never attested in the input but can be located in the space of possible grammars, including those used in faraway linguistic communities around the world (Crain and Thornton 2000; Yang 2002, 2006a). But the innate endowment of Universal Grammar cannot encompass the totality of linguistic knowledge that children must acquire. Language is full of idiosyncrasies that are remnants of history (Chomsky and Halle 1968): even the most hardened nativist would not suggest that the English tense *-d* rule is encoded somewhere in the genome to be awakened by the first instance of *walk-walked*. Language acquisition, then, must include an inductive learning process by which children form broad generalizations on the basis of examples — there is no escape from the indeterminacy of inference.

In the current study, I focus on an aspect of linguistic knowledge that poses special challenges for children: the acquisition of negative constraints. To learn a language is to learn its full range of possible expressions while never straying into the realm of the impossible. The first case study asks a simple question: Why can't we say **the asleep cat*? English has an interesting set of adjectives, the so-called a-*adjectives*, all of which start with a schwa (e.g., *afraid*, *alone*, *asleep*, *away* etc.) and all resist attributive usage in a prenominal position as in **the asleep cat*. The acquisition of the *a*-adjectives turns on the more general question: How do we learn what *not* to say? The second case study is the well-known problem of Baker's Paradox (Baker 1979), which has featured prominently in the learnability and development literature. Again, the problem concerns negative knowledge of linguistic forms. Consider the two dative constructions in (1a) and (1b):

(1)   a. John told the story to Bill.
          John told Bill the story.
     b. Mommy promised a cake to Mauve.
          Mommy promised Mauve a cake.
     c. John donated the painting to the museum.
          *John donated the museum the painting.

These verbs can freely alternate between the double-object and the *to*-dative construction. What prevents children from generalizing this pattern of interchangeability to verbs such as *donate* as in (1c)? Like the case of the *a*-adjectives, the dative constructions are also highly language specific. In Korean (e.g., Jung and Miyagawa 2004), the equivalent of the double-object construction is restricted to a handful of verbs, and there are languages such as Chamorro that disallow the double-object construction altogether (Chung 1998; Cooreman 1987; Topping 1973). Again, inductive learning from data is inevitable.

A typical, and very tempting, approach to these inference puzzles is to appeal to indirect negative evidence. I advise against this move. Indirect negative evidence is far too powerful and complex as a general learning strategy for language, and its side effects outweigh whatever benefits it brings to learners.

6.1.2 Indirect Negative Evidence

Put succinctly, indirect negative evidence means that the absence of evidence is the evidence of absence: the failure to observe certain forms in the input implies that such forms are ungrammatical. This of course doesn't follow as a matter of logic. But a logically flawed inference may nevertheless pass for a psychological principle of learning (Tenenbaum and Griffiths 2001): the human mind, a biological object, does not have to live up to philosophical standards. And it's easy to see why indirect negative evidence comes in handy for the acquisition of negative knowledge: if learners consistently fail to observe pattern such as *the asleep cat* in the input — the absence of evidence — they may conclude that these expressions are ungrammatical — the evidence of absence.

In the context of language acquisition, indirect negative evidence has been invoked mostly to deal with the Subset Problem (Angluin 1980; Berwick 1985; Chomsky 1981). Suppose, as depicted in Figure 6.1, the target hypothesis is g but the child has conjectured a superset hypothesis G instead. The learner will never be contradicted by the learning data because every instance of g is also compatible with the more general G. Indirect negative evidence to the rescue: if learners fail to observe "+" forms which are expected under G but not g, then they can retreat back to g in response.

Indirect negative evidence comes in many shapes and forms; see Pinker (1989) for a review. At its core, indirect negative evidence is the stand-in for direct negative evidence that would have been highly effective in language

6.1 Inference and Weight of Evidence

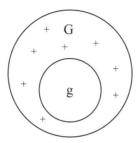

Figure 6.1
The target, and smaller, hypothesis g is a proper subset of the larger hypothesis G.

acquisition (Gold 1967). Consider again how learners may retreat from the superset grammar (G) to the subset grammar (g). If gives some negative feedback is given when learners produce a "+" expression in Figure 6.1 — "You can't say that!" — then presumably they can back off to the smaller g or at least know the larger G is flawed. But by now it is very clear that negative evidence is not generally available, or has little effect even if parents get involved in the business of language teaching (Bowerman 1988; Braine 1971; Brown and Hanlon 1970; Marcus 1993). The cross-cultural study of linguistic interaction between children and adults further suggests that negative evidence is unnecessary for language acquisition to succeed (Heath 1983; Peters 1983; Schieffelin and Ochs 1986).

Berwick's (1985) Subset Principle approaches the Subset Problem by requiring learners to attend to the smaller/subset hypothesis first. But the computational complexity of using the Subset Principle is too high to be practical. In order to determine the subset-superset relation between G and g, learners may need to compare the extensions of these hypotheses (i.e., the potentially infinite sets of strings they generate), which in the general case is not even computable (Osherson et al. 1986). In a study of a linguistically realistic domain of syntactic parameters, Fodor and Sakas (2005) find that the computational cost of detecting subset relations among a finite set of grammars is prohibitively high. Without a feasible means of computing the expectations of hypotheses — that is, the "+" expressions in the nonoverlapping region of $(G-g)$ — indirect negative evidence is unusable.

Probabilistic learning models such as Bayesian inference provides a natural formulation of indirect negative evidence. Failing to observe a sentence in a corpus of one thousand words is one thing — one may not have got around to use it — but its absence in a corpus of fifty million words is more conspicuous and thus constitutes a stronger cue for ungrammaticality. For concreteness, consider a typical model of Bayesian inference, which consists of the following components:

(2) a. A set of hypotheses to be selected H;
b. The *prior* probabilities of the hypotheses $P(H)$;
c. A likelihood function that calculates the probability of the observed learning data D under the hypotheses H, or $P(D|H)$, to obtain
d. The *posterior* probabilities of the hypotheses $P(H|D) = P(H)P(D|H)/P(D)$ (Bayes's rule), which are normalized by $P(D)$ to select the optimal hypothesis $\hat{H} = \operatorname*{argmax}_{H} P(H)P(D|H)$.

Consider a nonlinguistic example. Suppose that there are two coins, a fair one and a biased one with both sides. We randomly select a coin and would like to know whether it is fair or biased on the basis of its behavior. Here the prior probabilities of $P(H = \text{fair})$ and $P(H = \text{biased})$ are both $1/2$. The data, D, consists of 100 trials of coin toss where we observed only heads. The likelihoods are, $P(D|H = \text{fair}) = 1/2^{100}$ and $P(D|H = \text{biased}) = 1$. Multiplying by the prior probabilities, it is clear than "$H = \text{biased}$" is overwhelmingly favored over "$H = \text{fair}$."

This is all very reasonable because the competing hypotheses are mutually exclusive: the one is either fair or biased but not both. Troubles arise when the Bayesian model inherits the trappings of indirect negative evidence where, crucially, the competing hypotheses form superset-subset relations and are thus not mutually exclusive. More specifically, the Bayesian approach requires a likelihood function that favors the subset grammar (g) over the superset grammar (G)—the so-called Size Principle (Xu and Tenenbaum 2007, 252), that "hypotheses with smaller extensions assign greater probability than do large hypotheses to the same data, and they assign exponentially greater probability as the number of consistent examples increases." So we are back to the intractable problem of comparing grammar extensions, and the challenges in the classical nonprobabilistic learning framework remain the same (see Niyogi 2006 for general discussion). For these and other reasons, most advocates of Bayesian learning and related approaches (e.g., Chater and Vitányi 2007; Feldman et al. 2013; Goldwater et al. 2009; Perfors et al. 2011; Xu and Tenenbaum 2007) explicitly disavow any claim of psychological mechanisms. Even so, Bayesian learning tends to be extremely difficult to implement. It is routine

6.1 Inference and Weight of Evidence

for models to consume hundreds of hours of computer time to process a few thousand English words (Frank et al. 2009; Sirts and Goldwater 2013) but are outperformed by much simpler learning models directly motivated by the behavioral study of language acquisition (Lignos 2010; Stevens et al. 2016). The direction taken by the Bayesian program does not seem to shed light on the language acquisition process; see Yang 2016 for a systematic evaluation with considerations from language acquisition, variation, and change. Section 6.2 below provides a true test for Bayesian learning under realistic conditions of language acquisition that involve specific linguistic details: even if we put the conceptual and computational issues asides, the Bayesian learning approach still fares badly when it attempts to incorporate indirect negative evidence.

6.1.3 Sufficiency of Positive Evidence

Instead of worrying about how to use indirect negative evidence, let's consider why one may need indirect negative evidence to begin with. The primary motivation for indirect negative evidence is to allow learners to retreat from overgeneralization — but who says they need to generalize so aggressively and right away? The desert-island example that initiated our discussion was meant to show that it is sometimes wise to withhold judgment pending further evidence. If learners do not (absurdly) conclude that all dinosaurs fly on seeing only one in flight (a pterosaur in a picture book), then there is no overgeneralization to retreat from. In other words, learners needn't jump to a conclusion, positive or negative, about the unseen data unless a sufficient amount of evidence has been accumulated to support valid inference.

Sufficiency, I propose, follows the Tolerance Principle:

(3) *Sufficiency Principle*

Let R be a generalization over N items, of which M items are attested to follow R. R can be extended to all N items if and only iff:

$$N - M < \theta_N \quad \text{where} \quad \theta_N := \frac{N}{\ln N}$$

Before the positive evidence is sufficient — that is, when M sits below the sufficiency threshold — learners lexicalize all M items and does not generalize beyond them. Without any kind of generalization, the problem of overgeneralization does not even arise and indirect negative evidence needn't apply. That is, Baker's Paradox involving the verb *donate* as in (1) is a nonissue unless children have observed a significant majority of similar verbs attested in the double-object and *to*-dative construction — which, as we will see, probably won't happen during the first few years of language acquisition. Only when M crosses the Sufficiency threshold does R become a truly productive rule.

It is easy to see that the Sufficiency Principle has a built-in mechanism for retreating from overgeneralization. Suppose $N = 50$ and a learner has finally accumulated $M = 40$ instances to warrant a generalization ($50 - 40 = 10 < \theta_{50} = 12$). But further down the road, N may start to increase again. Suppose that N stands at 60 but no additional growth of M has been observed (still 40): now the rule will cease to be productive ($60 - 40 = 20 > \theta_{60} = 14$) and the learner will lexicalize all 40 items, so the once-productive generalization will be abandoned. The dynamics of learning under the Sufficiency Principle are exactly the same as under the Tolerance Principle, which was extensively covered in the preceding pages.

Although the formalisms for the Tolerance and Sufficiency Principle are similar, an important logical and empirical difference remains. The Tolerance Principle keeps track of exceptions to a rule R (i.e., attested items that explicitly defy R). The Sufficiency Principle, by contrast, asserts that unless the Sufficiency threshold has been crossed, learners are in a state of ambivalence regarding the $(N - M)$ items with which they have no direct experience: "I don't know" is an acceptable answer. That is, in contrast to the use of indirect negative evidence, the Sufficiency Principle does not conclude that unattested forms are ungrammatical — or grammatical, for that matter.

This may seem like a radical departure from traditional language learnability research but I cannot think of a single instance of language learning that does *not* require a sufficient body of evidence. Word learning is the most obvious example. Children's vocabulary acquisition is remarkably rapid and accurate but they usually cannot learn the meaning of a word right away. Granted, a landmark study by Carey and Bartlett (1978) shows that young children are capable of "fast mapping": they can recall the label for a novel object after very brief prior exposure. While clearly a critical ingredient in word learning, fast mapping should not be equated with successful word acquisition. First, children's performance was far from perfect in the original Carey and Bartlett study. Second, fast mapping has always been established in a forced-choice task, which is widely used in word-learning experiments. But being able to choose one meaning out of a finite (and small) set of candidates is not quite the same as *knowing* the word — that is, using it appropriately — as anyone who has taken a multiple choice exam must be aware. In fact, word learning must require repeated exposure, at least in the general case. For one thing, children are already very impressive word learners, averaging about ten words a day (Miller 1991); if word learning were truly "one shot," children would presumably attain college-level vocabulary in no time. For another, in a series of experiments, subjects (both children and adults) are exposed to novel words in a sequence of observations. Although they generally identify the target words better than the distractors, they are nevertheless far from perfect (Medina et al. 2011; Smith and Yu 2008; Stevens et al. 2016). Given the enormously high level of ambiguity in word learning tasks (Chomsky 1959; Gillette et al. 1999; Gleitman 1990; Landau and Gleitman 1984), children must be cautious learners that rely on the accumulation of confirmatory evidence otherwise they would not be able to achieve accurate vocabulary acquisition (Dromi 1987; Huttenlocher and Smiley 1987; Rescorla 1980). Presumably they don't use a word until they are fairly confident about its meaning, which requires cumulative evidence.

The acquisition of phonological structure also relies on the quantity of positive information. For example, native speakers have a strong sense of what constitutes a possible phonological word in their language (Halle 1978, 294):

(4) ptak thole hlad plast sram mgla vlas flitch dnom rtut

While none of these forms are actual English words, *thole*, *plast* and *flitch* are potential words (that may appear in the future) while the others stand no chance. There is a long line of research on the nature and acquisition of phonotactics, especially the onset and coda clusters permissible in a language (Coleman and Pierrehumbert 1997; Hayes and Wilson 2008; Scholes 1966).[1] Developmental studies clearly show a frequency effect for phonotactics in speech perception and production. For example, nine-month-old American infants prefer listening to words that follow the phonotactics of English rather than Dutch, despite the segmental similarities between these languages (Jusczyk, Houston, and Newsome 1999). Young children's production of consonant clusters also appears correlated with their frequency of occurrence in the input language (Coady and Aslin 2004).

---

1 A possibility remains that the phonotactics are epiphenomenal (Chomsky and Halle 1968). If so, the well-formedness of onsets and codas in words such as those in (4) would be reflexes of the phonological properties of the words in the language, rather than an independent component of grammatical knowledge (say, a list of permissible clusters). See Gorman 2013 for a recent and quantitative development of this position.

The acquisition of syntax most clearly demonstrates the cumulative effect of evidence. The premise of the variational learning model (Yang 2002) is to reject categorical models of parameter setting (e.g., triggering; Gibson and Wexler 1994) where a single instance of informative data can lead to radical revisions of the grammar. I will not review the extensive evidence for the gradualistic and against the discrete conception of parameter setting but will return to these matters briefly in chapter 7.

In what follows, I use the Sufficiency Principle to model children's grammatical development. The acquisition of a-adjectives is very straightforward (section 6.2) while the case of dative constructions is considerably more complex (section 6.3). As we will see, the validity of dative generalization fluctuates as the positive evidence in the input is sampled, much like the cases of phonological and morphological acquisition explored in chapter 4.

6.2 Why Are There No Asleep Cats?

A theory of language acquisition should provide a broad account of the speaker's linguistic competence. It must explain the expressions speakers can produce as well as the absence of expressions that they have not produced and in fact cannot produce, for these expressions are prohibited by universal or language-specific constraints.

Consider the so-called a-adjectives in this context. It has been observed (Beard 1995; Bolinger 1971; Bouldin 1990; Cinque 2010; Huddleston and Pullum 2001; Larson and Marušič 2004) that a class of English adjectives can only be used predicatively but not attributively in a prenominal position. These adjectives start with an unstressed schwa (a; hence the collective label a-adjectives):

(5) a. The cat is asleep. ??The asleep cat.
   b. The boss is away. ??The away boss.
   c. The dog is awake. ??The awake dog.
   d. The child is alone. ??The alone child.
   e. The troops are around. ??The around troops.

6.2.1 The Failure of Indirect Negative Evidence

The *a*-adjectives offer a perfect application for indirect negative evidence. The learner is to acquire a subset hypothesis (g) which only allows predicative usage, and to reject a superset hypothesis (G) which admits both predicative and attributive usage, which will allocate some probability mass for patterns such as "the asleep cat". For a corpus of English data (D) that contains *a*-adjectives, the expected attributive usage under G fails to show, thereby gradually lowering the likelihood of $P(D|G)$. Given enough time, the posteri probability $P(G|D)$ will lose out to $P(g|D)$, allowing the learner to adopt the correct, subset, hypothesis g. A search in the roughly two million words of child English in the public domain (MacWhinney 2000) yields about 2,300 tokens of *a*-adjectives: not a single instance of attributive usage is found. Although the child data is pooled from a large number of subjects, the average age of the learner is just over 2;10. English-learning children evidently acquire the syntactic properties of *a*-adjectives very early, and the distributional evidence must be robustly available.[2]

Previous accounts of a-adjective acquisition follow exactly this argument even though they are not always explicitly formulated in a Bayesian framework. For frequency-based accounts such as Stefanowitsch (2008), if a sufficiently frequent adjective fails to appear in an attributive position, it would constitute as evidence for its ungrammaticality. Similarly, Boyd and Goldberg (2011) invoke the strategy of *statistical preemption*: instead of using non-occurrence as cues for ungrammaticality, the child assumes that the syntactic forms that realize the same meaning are mutually exclusive: as Pinker (1989) notes, this is also a form of indirect negative evidence as suggested by Wexler and Culicover (1980) and Clark (1987). More concretely, if the semantic expression "the asleep cat" is called for, and the child consistently observes attestations in the relative clause (e.g., "the cat that is asleep"), then they will over time conclude that the attributive form is impossible. That is, the relative clause form preempts the attributive form.

It is easy to cast these proposals in a Bayesian formulation. In fact, recent proposals of Bayesian learning have made use of *overhypothesis* (Kemp et al. 2007), adopted from Goodman's well-known discussion (1955), a form of knowledge abstracted over a class of individual items. Hierarchical Bayesian

---

2 It is logically possible, though I believe unlikely, that the superset hypothesis of attributive usage is already ruled before the child has uttered a word. If so, then the onus is on the advocate of indirect negative evidence to demonstrate the reality of this very brief stage.

models can make inference over multiple levels of abstraction, from individual words and sentences to lexical casses and rules to universal constraints on language; see Perfors et al. 2010 for an application in syntactic learning. In the present study, one may wish to consider the a-adjectives as a natural class, and there are in fact linguistic and developmental evidence that children must do so; see section 6.2.2 and section 6.2.3 for discussion. If so, the indirect negative evidence would be stronger and presumably more effective: the Size Principle will disfavor the superset hypothesis more significantly because "the number of consistent examples" (Xu and Tenenbaum 2007) is larger when evaluated on the entire class than on any specific lexical item.

But these proposals do not seem to work, whether or not the a-adjectives as evaluated as individual items or collectively as a class under the overhypothesis formulation. The fundamental problem can be started simply: the superset hypothesis cannot be effectively ruled out due to the statistical properties of child-directed English. As such, indirect negative evidence leads to very poor learning results such that the a-adjectives cannot be distinguished from typical adjectives.

Our empirical study draws from two sets of data from the public domain. The first part is a parsed corpus of approximately 180,000 child directed sentences about 440,000 words in all (Pearl and Sprouse 2013). The parsed corpus facilitates search for specific syntactic structures that will be important for evaluating adjectives. The second part is the five-million-word corpus of child-directed English used in previous chapters. Both sets of data contain exactly twelve a-adjectives:

(6) across, afraid, ahead, alike, alone, apart, around, ashamed, asleep, awake, aware, away

Consider the use of indirect negative evidence that exploits frequency and absence. All the twelve a-adjectives are relatively common words so as to appear in a modest 440,000 word corpus. But none of them is sufficiently frequent such that their failure to appear attributively would be remarkable. In the parsed corpus, there are 517 predicatively used adjectives, including the twelve a-adjectives, with an average frequency of 13.75. There are also 575 attributively used adjectives with a noun phrase, with an average frequency of 14.73: the a-adjectives, as expected, never appear there. The intersection of the two sets produces 198 adjectives that are used both predicatively and attributively, with an average frequency of 57.7. This is also expected, because higher frequency adjectives have more opportunities to be used in both constructions. But only one of the 12 *a*-adjectives (*afraid*, with a frequency of 73 out of 440,000) falls into this higher frequency range; and many of the other 11 *a*-adjectives appear only once or twice and their absence of attributive use is not at all conspicuous. At the same time, even a cursory search reveals that the corpus contains many typical adjectives (e.g., *careful*, *sorry*, *ready*) that are much more frequent than *afraid* but appear exclusively predicatively: unlike the *a*-adjectives, these adjectives *can* appear attributively.

Evaluating the *a*-adjectives as a class under the overhypothesis approach (Kemp et al. 2007; Perfors et al. 2010) does not help. In the parsed corpus, there are collectively 143 *a*-adjectives. But even as a class, the *a*-adjectives are still less frequent than typical adjectives such as *sorry* and *careful*, which do not appear attributively: thus the Size Principle cannot accurately identify the *a*-adjectives as a class either. In fact, even in the five-million-word corpus, a part-of-speech tagger shows that the *a*-adjectives are, even collectively, still less frequent than *sorry* and *careful* which still do not appear attributively.[3] Even if one were to abandon the rigidity of logical inference, the absence of evidence still cannot be used as evidence of absence.

The strategy of statistical preemption by the relative clause paraphrase (Boyd and Goldberg 2011) fares far worse. I direct the reader to Yang 2015 for details. The problem is that adjectives are very rarely used in relative clauses to modify noun phrases (on average, only once almost every 3,000 utterances). Only three out of the twelve *a*-adjectives are used in a relative clause at all in five million words of child-directed English, and there are also many typical adjectives that, when modifying noun phrases, are exclusively used in the relative clause form. The rates of false positives and false negatives under the paraphrase preemption are extremely high.

Taken together, it is very unlikely for indirect negative evidence to reveal the syntactic properties of the *a*-adjectives. Once the superset hypothesis (attributive plus predicative usage) is introduced, there is no sufficient statistical evidence to rule it out. The failure of indirect negative evidence can be attributed to the inherent statistical distribution of language. Under Zipf's law, which applies to linguistic units (e.g., words) as well as their combinations (e.g., N-grams, phrases, rules; see section 2.1), it is very difficult to distinguish low probability events and impossible events. In the present case, the statistical

---

3 Presumably, there are more than two such typical adjectives, but here a cursory evaluation is sufficient.

distribution of language cannot separate the *a*-adjectives that resist attributive usage by design from the typical adjectives that fail to show attributive usage by chance.

The correct identification of *a*-adjectives, which I develop in detail elsewhere (Yang 2015), is to turn indirect negative evidence on its head. Note that the superset hypothesis is defined in terms of the distributional *differences* between *a*-adjectives and typical adjectives, but these differences are undetectable in realistic linguistic input as we have just seen. The alternative strategy is a positive one, as it exploits the distributional *similarities* between *a*-adjectives and other linguistic units, the latter of which resist attributive usage for independent reasons. Thus, the acquisition of *a*-adjectives becomes a problem of reduction, with an additional step of inductive generalization sanctioned by the Sufficiency Principle.

6.2.2 *A*-adjectives Are Not Atypical

Historically, many of the *a*-adjectives were derived from prepositional phrases Long (1969). While the etymology of words is unlikely to be available to learners, there is still synchronic evidence (Bruening 2011a,b; Rauh 1993; Stvan 1998) that reveals the PP-like characteristics of the *a*-adjectives.

First, locative particles *present, out, over, on/off, up, here/there* and so on are words that, like *a*-adjectives, also resist attributive use in a prenominal position:

(7)   a. The batter is up. ??The up batter.
       b. The matches are over. ??The over matches.
       c. The delivery is here. ??The here delivery.

In this regard, both *a*-adjectives and locative particles pattern like prepositional phrases:

(8)   a. The ball is out of sight. ??The out of sight ball.
       b. The dog is behind the fence. ??The behind the fence dog.
       c. The singers are at ease. ??The at ease singers.
       d. The marbles are in the jar. ??The in the jar marbles.

Finally, as noted by Salkoff (1983, 299) and Coppock (2008, 181), *a*-adjectives share a well-defined morphological structure; they are not an arbitrary list of adjectives that happen to share an initial schwa. Indeed, the ungrammaticality

6.2 Why Are There No Asleep Cats?

of attributive usage appears associated not with the *a*-adjectives per se but with the aspectual prefix *a*-, as shown in the novel adjectives (*abud* and *afizz*) below:

(9) a. The tree is abud with green shoots.
??An abud tree is a beautiful thing to see.
b. The water is afizz with bubbles.
??The afizz water was everywhere.

In a similar vein, Larson and Marušič (2004) observe that all *a*-adjectives can be decomposed into the prefix *a*- and a stem that is typically free but sometimes bound (e.g., *aghast* with *ghast* appearing in *ghastly*, *afraid* with *fraid* in *fraidy*, *aware* with *ware* in *beware*). The following list is taken from their paper with a few of my own additions; all resist attributive usage:

(10) abeam, ablaze, abloom, abuzz, across, adrift, afire, aflame, afraid, agape, aghast, agleam, aglitter, aglow, aground, ahead, ajar, akin, alight, alike, alive, alone, amiss, amok, amuck, apart, aplenty, around, ashamed, ashore, askew, aslant, asleep, astern, astir, atilt, aware, awash, away, awhirl

By contrast, the attributive restriction disappears if the adjective consists of the schwa *a*- (i.e., nonprefix) and a nonstem (11a) or a pseudostem (11b):

(11) a. The above examples
The aloof professor
The alert student
The astute investor
b. The acute problem

Of course, the language acquisition problem does not go away under the morphological approach to *a*-adjectives. First, learners need to recognize that the *a*-stem combination forms a well-defined set of adjectives structurally distinct from the phonologically similar but morphologically simplex adjectives. Second, and more importantly, they need to learn that the *a*-adjectives thus formed cannot be used attributively, which is the main problem at hand.

I will not review all the similarities between *a*-adjectives and locative participles discussed by previous researchers. Some of the diagnostics, such as those based on semantic considerations and -*ly* suffixation to establish the syntactic categories of *a*-adjectives, have proved less than conclusive (e.g., Bruening

2011a; Goldberg 2011). More critically, these diagnostics use *ungrammatical* examples: while invaluable to theorists in uncovering the complexity of linguistic knowledge, they are of no help to language-learning children. Moreover, attested examples from the web or large electronic corpora, which previous researchers used extensively to study the properties of the *a*-adjectives, cannot be assumed to be available to learners without further confirmation using naturalistic acquisition data.

However, one diagnostic has proved very robust and, as we will see, is abundantly attested in the input. A class of adverbs such as *right, well, far, straight* and so on, which expresses the meaning of intensity or immediacy, can be used to modify *a*-adjectives. Following Bruening 2011a, I collectively refer to these structures as *right*-type modification.

(12)   a. The coffee shop is right ahead.
       b. The baby fell right/sound asleep.
       c. The guards are well aware (of the danger).
       d. I was well/wide awake at four o'clock in the morning.

To be sure, probably not all *a*-adjectives may be used with *right*-type modification: *I became well/right afraid* is unacceptable to most English speakers I surveyed although one can find attested examples in very large corpora. But such adverbial modification cannot appear at all with typical adjectives (13a–b), while it is compatible with both locative particles (13c) and PPs (13d).[4]

(13)   a. *The car is right/straight/well new/nice/red.
       b. *The politician is right/straight/well annoying/amazing/available.
       c. The referee was right here/there.
          The cat ran straight out.
          The answer was wide off.
          The arrow was shot well over.
          The ball sailed far out.

---

4 It has been brought to my attention that *well, right*, and similar adverbs may appear with certain typical adjectives in British English and some varieties of Southern American English: examples such as "right nice," "well tired," etc. are grammatical and can be found on the web. It would be interesting to investigate additional features of these adverbs in such dialects—for example, the restrictions on the type of adjectives they can modify. Assuming that these speakers also disallow **asleep cats*, the morphological condition in (10) that sets the *a*-adjectives apart from typical adjectives must also play a crucial role in language acquisition.

6.2 Why Are There No Asleep Cats? 187

    d. The referee was right in the penalty box.
       The cat ran straight out of the house.
       The answer was wide of the mark.
       The arrow was shot well over the fence.
       The ball sailed far out of the park.

Despite these distributional similarities, I am somewhat reluctant to label *a*-adjectives as PPs or locative particles. PPs are, of course, phrasal while *a*-adjectives are single words. Locative particles are single words but they are morphologically simplex while also appearing to form a closed list. By contrast, *a*-adjectives are morphologically well structured and appear to be open-ended as illustrated in (9). Ultimately, the central issue is not what theorists decide to label the *a*-adjectives but how children acquire their syntactic properties on the basis of linguistic evidence.

6.2.3 Generalization with Sufficient Evidence

Our general strategy for learning the *a*-adjectives can be summarized as follows. Learners, on the basis of positive evidence, observe that *a*-adjectives pattern distributionally like locative particles and prepositional phrases; the resistance to attributive usage in the latter classes of linguistic units can be extended to the *a*-adjectives. However, not all *a*-adjectives are used in a PP-like context in the child-directed input; the Sufficiency Principle enables generalization across the entire class on the basis of sufficient positive evidence.

I first extracted the attested *a*-adjectives from the 5-million-word corpus of child-directed English. To identify the morphological structure of the *a*-adjectives, learners need to discover the compositional formation of *a*- with a stem. Morpheme segmentation takes place at a very early age (Gerken and McIntosh 1993) and children's morphological knowledge across languages is generally highly reliable (chapter 2). English learner's word segmentation errors (Brown 1973; Lignos 2011; Peters 1983; Yang 2004) such as *There are three dults in our family* (*dult* from *a-dult*) and *I was have* (/heiv/ as in *be-have*) further suggest that children recognize affixlike elements as phonological units during lexical acquisition and pursue them aggressively if the resulting materials could conceivably be a word. The morphological segmentation of *a*-adjectives would be further facilitated by the fact that the *a*-adjective stems are highly frequent and thus very likely to be part of a young child's vocabulary. For the present quantitative analysis, I processed the corpus data with a part-of-speech tagger to extract all the adjectives. I then segmented off a word-initial unstressed schwa *a* without violating the phonotactics of English; the results are presented here:

(14)   a. Containing stems: across, afraid, ahead, alike, alone, apart, around, ashamed, asleep, awake, aware, away
       b. Not containing stems: aberrant, above, acceptable, adept, adorable, affectionate, agreeable, allergic, amazing, American, annoying, another, approachable, attractive, available

Three distributional patterns emerge from the child-directed corpus. First, the presence or absence of a stem partitions the adjectives into two completely disjoint classes. All the items in (14a) contain a stem: all are *a*-adjectives. None of the items in (14b) contain a stem; all can be used attributively. Thus, the morphological criterion unambiguously defines the membership for the *a*-adjectives. Second, I searched for *right*-type modification usage for the non-*a*-adjectives in (14b): not a single instance was found, which supports the use of *right*-type modification as a diagnostic for distinguishing *a*-adjectives from non-*a*-adjectives. Third, eight out of the twelve *a*-adjectives in (14a) show robust usage with *right*-type modification. A corpus example for each *a*-adjective is provided in (15):

(15)   Are you wide awake?
       I'm well aware of my shortcomings thank you.
       Go right ahead.
       It fell right apart on you.
       Turn right around.
       Finish the book right away.
       He fell fast asleep.
       We are coming right across.

The attested examples range from a handful (for *apart*) to over a hundred (for *away*). While the child-directed corpus is not trivial in size, positive evidence for the distributional properties of *a*-adjectives will be even more robust in larger, and more realistic, samples of the primary linguistic data. I further note that the corpus contains numerous instances of *right*-type modification with locative particles as well as prepositional phrases: *right here* and *right there* appear over 3,000 times, *right up/over/on* are in the hundreds, *right off/-down/under* dozens, and so on. Therefore, a typical English-learning child will

6.2 Why Are There No Asleep Cats?

have plenty of opportunities to observe that *a*-adjectives, locative particles, and prepositional phrases are distributionally similar on the basis of positive evidence.

We are not quite done. Learners still need to form generalizations about the *a*-adjectives as a class: after all, only eight out of the twelve members of the *a*-adjectives in the input (14a) are used with *right*-type modification. But as the distributional sparsity of language shows (chapter 2), this is the rather typical situation in language acquisition. In almost all cases of language learning, children will not be able to witness the entire range of syntactic behavior for every member of a linguistic class. A case in point: there are forty-one *a*-adjectives in (10) that prohibit attributive use. Yet only twelve are attested in roughly a year's worth of child-directed English, and only twenty-eight appear at all in a fifty-one-million-word spoken American English corpus (SUBTLEX-US; Brysbaert and New 2009). Moreover, we can be fairly certain that not all, perhaps only a minority of them, will be used with *right*-type modification, the signature evidence that relates *a*-adjectives to locative particles and prepositional phrases. To the extent that native speakers resist the attributive use of the *a*-adjectives, they must have acquired this property on a relatively small but high-frequency set of examples. In fact, the reasoning here suggests that the acquisition of language is *only* possible with a relatively small, and "representative," amount of linguistic data; I return to this surprising point in chapter 7.

To summarize the quantitative analysis so far:

(16) a. On the basis of morphology, children identify a set $N = 12$ adjective-like words composed of a schwa and a stem.
b. $M = 8$ members of N are frequently modified by special classes of adverbs (e.g., *right*).
c. Locative particles (e.g., *here* and *up*) and prepositional phrases are also frequently modified by the same set of adverbs.

The Sufficiency Principle facilitates the very last step of the inference: Should children generalize a pattern observed for a subset of $M = 8$ items to the entire class of $N = 12$? The Sufficiency Principle suggests yes, and the decision again boils down to two numbers: 8 positive instances are enough ($12 - 8 = 4 < \theta_{12} = 5$). That is, children will assume that the properties of the 8 positively attested members of the *a*-adjective classes will extend to *afraid, ashamed, alone*, and *alike*, as well as to novel *a*-adjectives that follow the same morphological structures. This ensures that the property associated with locative particles, including the prohibition on attributive use, can productively hold for the entire class of *a*-adjectives.

I emphasize that the learning model developed here is most conservative. It is possible that children have access to other sources of information which provide more direct constraints against the prenominal use of the *a*-adjectives. For instance, the syntax and semantics of *a*-adjectives require additional research: perhaps there are other cues that help children identify the special properties of the *a*-adjectives. If so, the learners' task would presumably be further simplified. In fact, my proposal deals with a worst case situation, that the children are to acquire the properties of the *a*-adjectives with the most basic application of distributional learning. The Sufficiency Principle provides the learner with the confidence of generalization: if it walks like a duck and quacks like a duck, it must be a duck.

6.3 Resolving Baker's Paradox

The dative constructions in English present a perfect case of inductive indeterminacy. Again, consider the examples in (17), which have become a focal point of research since a classic paper by C. L. Baker (1979):

(17) a. John threw the ball to Bill.
      John threw Bill the ball.
   b. John told Bill the story.
      John told the story to Bill.
   c. John whispered the news to Bill.
      *John whispered Bill the news.
   d. John donated the painting to the museum.
      *John donated the museum the painting.

Examples such as (17a) and (17b) would seem to suggest that the double-object construction and the *to*-dative construction are mutually interchangeable, only to be disconfirmed by the ungrammaticality of *whisper* (17c) and *donate* (17d) in the double-object form. How do children avoid these traps of false generalizations?

6.3 Resolving Baker's Paradox

This long and complex case study is structured as follows. Section 6.3.1 reviews the basic constraints on dative constructions: the double object construction requires the semantics of caused possession and the *to*-dative construction requires the semantics of caused motion. Both constructions are productive in English but they take children a few years to acquire, with overgeneration errors littered along the way. Section 6.3.2 presents the distribution of the two constructions in the child-directed corpus. I show that the structural constraints on these constructions can be inductively learned from the set of attested verbs and thus needn't be stated as built-in features of Universal Grammar. Furthermore, I show that the Sufficiency Principle can derive the productivity of these constructions in English, because a sufficiently large number of verbs with the appropriate semantics are attested in the child-directed corpus. Section 6.3.3 tackles Baker's Paradox in its full glory. Latinate verbs such as *donate*, the source of the learnability puzzle, are hardly present in the child-directed corpus, making it impossible for learners to acquire the finer restrictions on the dative constructions. I show that by expanding the English data in some appropriate fashion, learners would have evidence to retreat from overgeneralizations under the Sufficiency Principle. However, productive subregularities can still be discovered by further partitioning the verbs, following the strategy employed for the acquisition of morphology and phonology in previous chapters.

6.3.1 Conditions on Dative Constructions

One solution to Baker's Paradox, a trivial one, is to claim that there is no generalization at play: children just remember what's the input and never go beyond that (Baker 1979; Fodor and Crain 1987). If adults never use *donate* and *whisper* in the double-object construction, neither will child learners. However, there is little doubt that both dative constructions are productive in English, and a strict version of lexical conservatism cannot be correct.

First, new verbs with certain semantic properties have entered the English lexicon and spontaneously participated in dative constructions. For the verb *text*, which became available in recent years via communication technology, both the double-object and the *to*-dative construction are possible: *John texted the news to Bill* and *John texted Bill the news*. Verbs like *snowboard*, surely a recent physical activity, can be used in the *to*-dative construction (*He snowboarded the medicine to the tourists trapped by the avalanche*). Ditto for *beam*, when used in the *Star Trek* sense (*Scotty beamed Captain Kirk to yet another planet*). Second, and more directly, there is converging evidence from child language that the dative constructions do extend beyond the input.

Gropen et al. (1989, 217) found a small but nontrivial number of naturally occurring dative errors in children's speech, which are quite easy to spot in the CHILDES corpus. The list in (18) gives a partial sample from that work, along with some additional examples reported in other studies:

(18)   a. From Gropen et al. (1989)'s corpus analysis of child speech
            You finished me lots of rings.
            Ursula, fix me a tiger.
            Jay said me no.
            Don't say me that. (Asking adult not to tell him to put on his socks)
            So don't please ... keep me a favor. (Asking brother not to throw up on a ride)

    b. From Bowerman and Croft (2008)
            Shall I whisper you something?
            Pick me up all these things.
            I said her no.
            You put me just bread and butter.
            Button me the rest. (Asking to fasten the open buttons of her pajamas)
            Choose me the ones that I can have.
            Mattia demonstrated me that yesterday.

While some of the examples may be regarded as benefactives (e.g., *Fix me a tiger*), examples such as *I said her no* are genuine double-object constructions, and genuine mistakes. Overall, 5% of the double-object tokens in the study by Gropen et al. are overgeneralizations, involving 14% of the verb types. These error rates are modest but not negligible: they are in fact comparable to that in overregularizations or irregular verbs, which ranges from from 4.2% (Marcus et al. 1992) to 8% (Maslen et al. 2004); see chapter 2. Both the past-tense rule of "-*d*" and the double-object construction must be regarded as productive.

Moreover, experimental studies have shown that when presented with novel verbs in one dative construction, children spontaneously extend them to the other. Gropen et al. (1989), working with seven-year-old children, find that the subjects are sensitive to a morphophonological constraint on the double-object construction (Green 1974; Oehrle 1976). Specifically, polysyllabic (and Latinate-like) nonce verbs are generalized less frequently (39.1%; *orgulate* and

6.3 Resolving Baker's Paradox

calimode) than monosyllabic nonce verbs (54.7%; *moop* and *keat*); see Randall 1980 and Mazurkewich 1984 for similar findings. Recent studies suggest that the productivity of the dative constructions emerges around the age of three. In Conwell and Demuth (2007)'s experiments, children observed a novel action that involved transferring an object from the child to a recipient via a conveyor belt or a catapult; the action was described as *pilk* or *gorp*. The experimenter modeled the use of the novel verb in one of the two dative constructions — *You pilked the cup to Toby* or *I pilked Petey the cup* — and the children were then asked to perform and describe these actions. Conwell and Demuth found that the children were capable of extending both constructions and were not limited to the usage form they were exposed to (contra Akhtar and Tomasello 1997). All in all, previous research on English datives suggests that both double-object and *to*-dative constructions are productive in child language: learners cannot be lexically conservative but must attack Baker's Paradox head on.

I do not have the space for an extensive review of the theoretical literature on the dative constructions. By now there is general agreement on what makes dative constructions possible (Beck and Johnson 2004; Goldberg 1995; Gropen et al. 1989; Hale and Keyser 2002; Krifka 1999; Levin 1993; Pinker 1999; Randall 1987).[5] For instance, Hovav and Levin (2008), Levin (2008), Bruening (2010) and others suggest that the double-object construction requires verbs of caused possession and the *to*-dative construction requires verbs of caused motion. The distinction between the two classes can be observed in the following examples (Gropen et al. 1989):

(19) a. I sent a package to the boarder/the border.
b. I sent the boarder/*the border a package.

Both *boarder* and *border* may be the destination of the motion caused by *sent*; thus both variants are grammatical for the *to*-dative construction (19a). By contrast, only a *boarder*, not *border*, may have the package as the result of caused possession, hence the grammaticality contrast in (19b). It has also been observed that Latinate verbs (e.g., *donate* as noted earlier) tend to resist the double-object construction. Language learners of course have no access to the etymology of the verbs, but as reviewed earlier, children appear sensitive to the morphophonological properties of words (e.g., polysyllabicity) associated with words of latinate origin. At the same time, it's worth stressing that these

---

5 I focus on grammaticality and put aside the issue of preference when both dative constructions are available; see Bresnan and Nikitina 2009 among others for discussion.

etymological and morphophonological constraints are only tendencies: Latinate verbs such as *award*, *assign*, and *promise*, which have similar semantics as *donate*, do participate in the double-object condition, while Germanic verbs such as *say* and *shout*, which are comparable to *tell*, do not.

The semantic properties of dative verbs are only *necessary* conditions on the availability of the constructions, or what Gropen et al. (1989) and Pinker (1989) call "broad-range rules." Presumably all languages have verbs of caused possession and caused motion, but whether a specific language allows the productive use of these constructions must be determined on the basis of experience. As reviewed earlier, there are languages where the equivalent of the double-object construction is not productive but limited to a handful of verbs (see Jelinek and Carnie 2003, Jung and Miyagawa 2004, and especially Harley 2002). Languages such as Chamorro lack the construction altogether (Chung 1998; Cooreman 1987; Topping 1973). Furthermore, it is evident that caused possession/motion are too coarse grained to account for the full range of facts regarding dative constructions: subgrouping of verbs is necessary via "narrow-range rules" (Gropen et al. 1989; Pinker 1989). Consider the contrast between *throw* and *whisper*:

(20) a. I threw the ball to him.
I threw him the ball.
b. I whispered the secret to him.
*I whispered him the secret.

According to to Gropen et al. (1989), the difference between *throw* and *whisper* is due to their membership in different narrow range rules/classes. Specifically, *throw* belongs to the class of "instantaneous causation of ballistic motion" (p. 243), which allows the double-object construction in English, whereas *whisper* falls in the class of "manner of speaking" (p. 244), which does not.

It should be clear that the broad- and narrow-range rules and other verb classifications are descriptions of Baker's Paradox, not solutions. It is highly unlikely that they are innate criteria for dative constructions, waiting to pick up verbs with prescribed properties. First, while the validity of the broad-range rules as a necessary condition for dative constructions is well supported across languages, the distribution of the narrow-range rules appears arbitrary. Table

6.1 summarizes a survey of narrow range-rules in several languages (from Levin 2008; see the references there).[6]

Table 6.1
Narrow-range classes and double-objection constructions across languages (ND = no data)

	Greek	English	Warlpiri	Hebrew	Icelandic	Mandarin	Yaqui	Fongbe
Give-type	Yes	Yes	Yes	Yes	Yes	Yes	Yes	Yes
Future having	Yes	Yes	ND	Yes	Yes	Yes	ND	ND
Send-type	Yes	Yes	Yes	Yes	Yes	No	No	ND
Bring/take	Yes	Yes	Yes	Yes/No	Yes	No	ND	ND
Throw-type	Yes	Yes	Yes	Yes	No	No	No	No
Push-type	No?	No?	No	No?	ND	No	ND	ND

Second, and even within a single language, the availability of dative constructions can be lexically arbitrary. Again, recall that the double-object construction is fine for *promise* but not *donate*, and the *to*-dative construction works with *tell* but not *ask*. Levin's (1993) encyclopedic study of verb classes gives us a sense of variability in English alone. For instance, Latinate verbs in (21a) are possible in the double-object construction while those in (21b) "belonging to some of the semantically plausibles classes" (Levin 1993, 46) are not.

(21) a. Available for double-object construction: advance, allocate, allot, assign, catapult, cede, concede, extend, grant, guarantee, issue, promise, refund, relay, render, rent, repay, serve

   b. Unavailable for double-object construction: address, administer, broadcast, contribute, convey, delegate, deliver, demonstrate, denounce, describe, dictate, dispatch, display, distribute, donate, elucidate, exhibit, explain, explicate, express, forfeit, illustrate, introduce, narrate, portray, proffer, recite, recommend, refer, reimburse, remit, restore, return, sacrifice, submit, surrender, transfer, transport

---

6 The *push*-type is given a question mark here for English, presumably reflecting inconsistent judgment reported in previous literature: Levin (1993, 46) treats *push* as allowing double objects but Gropen et al. (1989, 244) reject it.

Not all the speakers I consulted agree with Levin's judgment, and that's exactly the point. The distribution of the dative constructions within and across language is too idiosyncratic to be entirely determined by prespecified internal factors in Universal Grammar: surely no one wishes to claim that the narrow-range rule for "verbs of instrument of communication," including *radio, email, telegraph,* etc. (Gropen et al. 1989, 244), is innately available. The acquisition of the dative constructions must have a significant component of inductive learning. If so, we naturally wonder, in the spirit of a minimalist approach, about the extent to which the role of an innate Universal Grammar can be reduced in the design of language.

The most detailed account of dative acquisition is still the theory put forward by Pinker (1989, chap. 6). He proposes that children are innately equipped with verb semantic primitives, described by a theory that draws eclectically from the earlier literature and most closely resembles the Lexical-Conceptual Structure approach developed by Grimshaw (1990), Jackendoff (1990b), and others. The semantic primitives, such as EVENT, STATE, PATH, THING, PROPERTY, etc. (Pinker 1989, 208), provide the features from which the learner may derive the conditions for the dative constructions. Pinker also assumes the innateness of *linking rules* that map semantic arguments to their syntactic structures. The acquisition of the dative constructions amounts to learning the semantic-class membership of verbs: the linking rules then provide the syntactic realization. The innateness of linking rules is somewhat suspect, given that languages have very different ways of realizing the dative constructions, a point that Pinker also acknowledges but does not address (1989, 281). And there have been criticisms of the innateness assumption from the developmental literature (e.g., Brooks and Tomasello 1999).

In my view, the most important aspect of Pinker's theory is that semantic classes can be constructed from verbs on the basis of their syntactic behavior. Both the broad- and narrow-range rules are to be discovered inductively. Broad-range rules are generalized in a top-down fashion: learners form abstractions over the set of verbs that participate in the same construction (e.g., double object). Narrow-range rules, by contrast, are constructed bottom up, applying only to the narrowest semantic class that contains the verb in attested usage. Suppose learners observe that both verb X and verb Y are used in the form of "X NP NP" and "Y NP NP." If, by hypothesis, the semantic primitives of X and Y are innately provided, then learners can construct semantic classes based on the features of X and Y, as well as the correspondence between such semantic classes and their syntactic realization (i.e., "*Verb* NP NP").

6.3 Resolving Baker's Paradox

As will become clear, my proposal is similar to Pinker's in that both regard inductive learning as an important component in the acquisition of dative constructions — that is, learners need to discover the distributional properties of verbs and their syntactic behavior. However, the main problem with Pinker's proposal is that it has no way of accounting for the overgeneralization of errors in children's datives such as those listed in (18), nor any mechanisms by which children may retreat from these errors.

In fact, and quite surprisingly, Pinker (1989, 292—295) dismisses the significance of overgeneralization errors, attributing them to performance or other factors not pertinent to the dative constructions.[7] If so, then Baker's Paradox, and the problem of how to retreat from generalization, do not arise. Pinker interprets an error like *Jay said me no* as a child mistaking the verb *say* for *tell*, which does allow double objects. But there are at least three problems with this position, in addition to the unconstrained interpretative freedom afforded by his account. First, while the error rates are low (5% of all production tokens; Gropen et al. 1989), they are not negligibly so and are in fact comparable to overregularization errors (Marcus et al. 1992). Both are an order of magnitude higher than overirregularization errors (section 2.3.2) for which a performance-based interpretation is more credible. Second, it seems unlikely that children will get extremely common verbs such as *say* and *tell* mixed up; some of the dative errors listed in (18) were produced by children who were five to seven years old and it's very difficult to imagine that they would get the semantics of these verbs wrong. Finally, and most importantly, Pinker's account provides no explanation for Baker's original problem: How do we learn that *I donated the museum a painting* is ungrammatical? There is no mechanism that prevents verbs such as *donate* from appearing in the double-object construction, because its semantics is very similar to that of verbs such as *advance*, *offer*, and *promise* — all of which allow the double-object construction (22) — and will presumably fall into the narrow range rule "future having" (Gropen et al. 1989, 244):

(22) a. They assigned the class two problem sets. (Latinate like *donate*)
   b. They offered the driver fifty dollars. (Stress-initial like *donate*)
   c. They promised the citizens clean water. (Both Latinate and stress-initial like *donate*)

---

[7] His treatment of causative and passive errors similarly relegates them to performance factors, and it is also unpersuasive; see Bowerman and Croft 2008.

Slapping on additional narrow-range labels, which is always a possibility for theorists, is a restatement of the facts and clearly does not help language learners. Once again, the broad- and narrow-range rules are descriptions of the terminal state of learning, not an explanation of how children get there. As for Baker's Paradox, we are back to square one.

6.3.2 How to Text Me a Message

Let's consider how the Sufficiency Principle can be applied to the acquisition of the dative constructions. Our main goal in this section is to account for the productive use of the two dative constructions in child language, and the fact that new verbs such as *text* became immediately available for the dative constructions on entering the English language as verbs in the mobile age. In section 6.3.3, I explore how the negative exceptions such as *say*, *whisper*, and *donate* may be acquired—that is, children retreat from overgeneralization.

Like Pinker (1989), I assume that learners are innately equipped with the semantic primitives, and that the semantic conditions for the syntactic constructions are inductively learned from the verbs that participate in the constructions. Unlike Pinker, however, I do not assume the innateness of the linking rules that map between semantic classes and syntactic forms, because of the considerable variation in the realization of datives across languages, and because of the lexically arbitrary choice of verbs within a language.

The learning process goes as follows: I will use the double-object construction for illustrative purposes but the same applies to *to*-datives and other constructions.

(23) a. A child learner observes a set of verbs $V_1, V_2, \ldots, V_M$ that participate in the syntactic form of "V NP NP."

b. The learner proceeds to inductively identify a semantic class C, over the verbs $V_1, V_2, \ldots, V_M$.

c. The learner identifies the total number of verbs $(N, N \geq M)$ that fit the structural description of C.

d. If $(N-M) < \theta_N$ then the learner extends the use of double objects to all members of C.

e. Otherwise the learner lexicalizes the M verbs as allowing double-objects but will not extend the construction to any other item.

6.3 Resolving Baker's Paradox

Note the similarities with the numerical approach to morphological productivity. As I have suggested, all inductive processes in language acquisition follow the Tolerance Principle.

6.3.2.1 Datives in the Input

To understand how children learn the productivity of the dative constructions, and how they may backtrack from them, I again turn to the child-directed English corpus. The first step is to extract all the verbs attested in the double-object and the *to*-dative constructions from which semantic generalizations can be drawn. It is impractical to go through every utterance. As a remedy, I focus on the verbs identified as candidates for the double-object and *to*-dative constructions according to Levin's (1993, 45-47) classification, which appears to be an exhaustive listing of all dativizable verbs in English.

I extracted all inflected forms of these verbs from the child-directed corpus and manually determined if a verb appears in the double-object or the *to*-dative construction at least once. Verbs that never had an opportunity to be used datively, as judged by the context, are excluded from analysis. For instance, *slip* may be used in the double-object construction (*He slipped the maître d' fifty bucks to get seated right away*) but all attested forms in the corpus are either intransitive (*He slipped on the ice*) or transitive (*Slip the button in*). Similarly, *yell* may appear in the *to*-dative (*The captain yelled instructions to the sailors*), but all forms of *yell* in the corpus are intransitive (e.g., *Stop yelling*).[8]

My search results revealed some minor inconsistencies in Levin's classification. For example, *ask*, *call*, and *guarantee* are listed as permitting the *to*-dative construction but none of the speakers I consulted agreed; in the child-directed corpus, all three verbs appeared exclusively in the double-object construction. Levin also lists *sing* as a *to*-dative-only verb in the "manner of speaking" class, but "Marilyn sang the President a birthday song" is clearly grammatical and there are many attested double object uses in the child-directed corpus (e.g., *Can you sing Mommy a song?*). These discrepancies are minor, but it is important to note again that I did not rely on Levin's grammaticality judgment in the analysis of the datives. I used her lists for their exhaustiveness. Whether a verb

---

8 Levin (1993, 46) assigns a question mark to the *drive* class (*barge*, *bus*, *fly*, *truck*, etc.) as double-object permitting but most speakers I consulted rejected this use, nor did I find any attested examples in the child-directed corpus. These verbs are excluded from my analysis.

participates in a construction or not is strictly based on its attested usage in the child-directed corpus. If it does not appear in a dative construction, then it is not included in the set of positive examples (i.e., M in the Sufficiency Principle). Generalization is formed strictly on the set of attested dative verbs (M) as a proportion of the appropriate candidate set (N) of which M forms a subset.

The search produced two lists of verbs that appeared in the double-object and *to*-dative construction respectively. The two lists have considerable overlap, because many verbs (e.g., *give*) can participate in both constructions. While most of the past research on datives, especially the acquisition literature, focuses on why verbs such as *donate* can only appear in the *to*-dative but not the double-object construction, there is also a class of verbs (Levin 1993, 47) that can appear only in the double-object but not the *to*-dative: *bill, consider, dub, elect, find*, etc. These are also included in my analysis.

The general approach is to see what kind of semantic generalizations may emerge from the attested use of dative constructions and to evaluate their productivity. For clarity of presentation, I group verbs by their classification according to Levin (1993), which in turn is partly based on the narrow-range rules proposed by Gropen et al. (1989). Again, I do *not* presume this to be the way child learners organize the dative verbs; as will be seen shortly, I argue that these conditions are in fact learnable from the data, on the basis of shared semantic properties of the verbs that participate in these constructions.[9]

6.3.2.2 From Necessity to Sufficiency

Consider first the verbs that participated in the double-object construction:

(24) a. Give Verbs ("verbs that inherently signify acts of giving"): feed, give, lend, loan, pass, pay, rent, sell, serve, trade

  b. Verbs of Future Having ("commitments that a person will have something at some later point"): assign, grant, guarantee, leave, offer, owe, promise

  c. BRING and TAKE ("verbs of continuous causation of accompanied motion in a deictically specified direction"): bring, take

  d. Send Verbs ("verbs of sending"): hand, mail, send

---

[9] See the introductory remarks in Levin 1993 on her research methodology for an insightful discussion of how linguists, and potentially children, approach the problem of semantic classification.

6.3 Resolving Baker's Paradox

e. Verbs of Throwing ("instantaneously causing ballistic motion"): blast, throw, toss

f. Verbs of Transfer of a Message ("verbs of types of communicated message [differentiated by something like 'illocutionary force']"): ask, read, quote, show, teach, tell, write

g. Verbs of Manner of Speaking: sing

h. Non-Alternating Double-Object Only (Levin 1993, 47): bet, call, charge, consider, cost, find, make, name, pronounce

In roughly one year of child-directed English, a "typical" learner will have observed 42 verb types in the double-object form.

Of the 42 verbs, almost all are clearly verbs of caused possession that involve the transfer of objects, entities, or abstract information. Only 4 verbs — *call*, *consider*, *name*, and *pronounce* — fall outside of this generalization. In (25), I list some examples of their attested usage in the child-directed corpus:

(25) a. We will call him a Turbo.
      b. This is considered a sting.
      c. Last time you named him Wolfie.
      d. I now pronounce you man and wife.

I suggest that these 4 examples are tolerable exceptions. Indeed, their double-object usage appears to have shared semantics as well, as all are performative verbs; if so, they form a "parallel" semantic generalization, as discussed in the treatment of recursive productivity (section 3.5.4). In any case, 4 out of 42 is a negligible level of noise ($\theta_{42} = 11$), and the child will notice that the double-object verbs almost always have the semantics of caused possession.

(26) double-object $\Longrightarrow$ caused possession

which converges with the current consensus view on the semantic requirement of the double-object construction (Bruening 2010; Hovav and Levin 2008; Levin 2008). This necessary condition on the double-objects, then, seems learnable on the basis of the primary linguistic data, assuming, of course, that the child learner is equipped with the primitives that facilitate semantic generalization.

Consider now the properties of the verbs that participate in *to*-dative constructions. Again, I follow Levin's classification for clarity of presentation:

(27) a. Give Verbs ("verbs that inherently signify acts of giving"): feed, give, lend, pass, pay, rent, sell, serve, trade

b. Verbs of Future Having ("commitments that a person will have something at some later point"): assign, award, grant, guarantee, leave, offer, promise

c. BRING and TAKE ("verbs of continuous causation of accompanied motion in a deictically specified direction"): bring, take

d. Verbs of Transfer of a Message ("verbs of types of communicated message [differentiated by something like 'illocutionary force']"), including the Say-class (*do*-dative only): mention, quote, read, report, say, show, teach, tell, write

e. Latinate (*to*-dative only): address, deliver, describe, donate, explain, introduce, return, transport

f. Send Verbs ("verbs of sending"): hand, mail, send, ship

g. Slide Verbs: bounce, roll

h. Carry Verbs ("verbs of continuous causation of accompanied motion in some manner"): carry, drag, haul, hoist, pull, push, shove, tow

i. Verbs of Throwing ("instantaneously causing ballistic motion"): flip, hit, kick, shoot, throw, toss

j. Verbs of Putting with a Specified Direction (*to*-dative only): drop, lift, raise

k. Verbs of Fulfilling ("X gives something to Y that Y deserves, needs, or is worthy of"; *to*-dative only): credit, present

l. Verbs of Manner of Speaking (*to*-dative only): shout, sing, whisper

Here learners observe 63 distinct verbs used in the *to*-dative construction. These turn out to be a heterogeneous group. It is difficult to provide a single coherent criterion to capture the semantics of these verbs, but it seems clear that both caused possession as in ((27a–e); *give, award, bring, donate,* etc.) and caused motion (27f–j); *hand, bounce, carry, throw, raise,* etc.) are well represented, with type frequency of 35 and 23 respectively. The verbs in (27k–l) are somewhat difficult to classify. The net effect is that neither class is small

6.3 Resolving Baker's Paradox

enough to be ignored, unlike the four oddball verbs in the case of the double-object construction in (24). A parallel application of the Tolerance Principle (section 3.5.4) suggests that learners formulate the generalization disjunctively.

(28) *to*-dative construction $\implies$ caused possession or caused motion

This recalls the analysis of German noun plurals (section 4.4). When the Tolerance Principle fails to discover a productive suffix for all nouns, learners partitions the nouns by gender, where a productive suffix emerges disjunctively for each subclass.

The analysis of (26) and (28) suggests that the semantic prerequisite for the dative constructions can be inductively acquired from the primary linguistic data. The directionality of inference can be turned around: after all, children produce over-generalization errors such as *Choose me the ones that I can have*, *Jay said me no*, etc. The Sufficiency Principle suggests that there must be large enough numbers of caused possession and caused motion verbs that *are* used, respectively, in the double-object and *to*-dative construction.

Thus, we need to look at the caused-possession verbs (N) in the child-directed input and count those actually used in the double-object construction (M): if M is sufficiently large compared to N, then children would be justified in turning the inductively constructed generalization in (26) into a productive rule, that any verb with the semantics of caused possession should allow the double-object construction.

We already know from (24) that the value of M is 38 as in (29a). Corpus search turns up 11 additional verbs that have the semantics of caused possession but did not appear in the double-object construction:

(29) caused possession $\overset{?}{\implies}$ double objects
   a. Attested in double object: ask, assign, bet, blast, bring, charge, cost, feed, find, give, grant, guarantee, hand, leave, lend, loan, mail, make, offer, owe, pass, pay, promise, quote, read, rent, sell, send, serve, show, sing, take, teach, tell, throw, toss, trade, write (38)
   b. Not attested in double object: address, deliver, describe, explain, introduce, mention, report, return, say, ship, transport (11)

The 11 verbs in (29b) make an interesting list. For verbs such as *introduce* and *say*, the double-object construction is ungrammatical:

(30) *John introduced the children a new dish.
       *John reported the police the crime.

*John said Bill something mean.

whereas others do allow the double-object construction but did not have the opportunity to do so (e.g., caretaker used the *to*-dative variant instead):

(31) John shipped Bill his purchase.

But of course, the children have no way of knowing *why* the verbs in (29b) fail to appear in the double-object construction (ungrammaticality or lack of opportunity); they can only draw a conclusion from the attested examples. Nevertheless, a sufficiently large number ($M = 38$) is able to justify the generalization over a set of $N = 49$: 37 ($49 - \theta_{49}$) would have been enough.

(32) caused possession $\Longrightarrow$ double-object

The justification for productivity in (32) is quite precarious: the positive examples barely cleared the threshold. But consider the following fact. The caused-possession verbs attested (24) in the double-object construction are statistically much more frequent ($p < 0.05$) than the unattested (29b) — by a ratio of almost 3:1 on average in the child-directed corpus. Only one of the ten most frequent caused-possession verbs, *say*, fails to show double-object usage. In other words, a sample of caused-possession verbs in child-directed speech is very likely to contain an overwhelming majority of those used in the double-object construction, facilitating the learning of the rule in (32).

Now we're immediately able to account for the overgeneralization errors such as *Jay said me no* and others reported in (18). These verbs fit the semantic criterion for the double-object construction — *say* is very similar to *tell* because both involve the transmission of information — which is sufficiently supported in the linguistic input. The children will eventually need to retreat from the productivity of (32), which I address in section 6.3.3.

The analysis of the *to*-dative construction is similar. First, caused possession overwhelmingly supports the *to*-dative construction:

(33) caused possession $\stackrel{?}{\Longrightarrow}$ *to*-dative

   a. Attested in *to*-dative: address, assign, award, bring, deliver, describe, donate, explain, feed, give, grant, guarantee, introduce, leave, lend, mention, offer, pass, pay, promise, quote, read, rent, report, return, say, sell, serve, show, take, teach, tell, trade, transport, write (35)

   b. Not attested in *to*-dative: ask (1)

6.3 Resolving Baker's Paradox

Virtually every caused-possession verb participates in the *to*-dative construction, well above the Sufficiency threshold. Thus, we have

(34) caused possession $\implies$ *to*-dative

Although previous research on the acquisition of dative constructions focuses mostly on the overgeneralization of double-objects, children also overgeneralize (34). I found one instance of *ask* in the *to*-dative form in the Hall Corpus in CHILDES.

(35) (Mother talking about needing to eat meat before cookies)
Child (4;6): When you gonna feed me? I asked this to you.

The use of *ask* here is quite striking. It marks the transfer of abstract information ("this" = "when you gonna feed me," the direct object) from the child to the mother (the indirect object), exactly the required semantics of caused possession, resulting in a quite unusual use of the verb *ask*.

Before concluding this section, I wish to offer a novel argument for the use of statistical evidence in theoretical linguistic analysis. As is well known, the syntactic nature of the double-object and the *to*-dative constructions have been extensively debated. For some authors (e.g., Baker 1988; Larson 1988, 1990, these two constructions are derivationally related; for others (e.g., Bruening 2010; Harley 2002; Jackendoff 1990a; Pesetsky 1995), they are separate constructions without implicational relations. A (strong) prediction of the derivational account would be that verbs that appear in one construction should be automatically extendable to other construction. This can be formalized quantitatively. Suppose that there are N lexical items that participate in construction A, out of which M also appear in construction B. Following the Sufficiency Principle, if $N - M < \theta_N$, then learners are justified to conclude that construction A and B are productively implicational.

Turning now to the two constructions, there are 23 caused-possession verbs that appear in both the double-object and the *to*-dative constructions; these are the intersection of the sets in (29a) and (33a). But 23 falls short of the Sufficiency threshold for both implicational direction. There are $N = 38$ verbs in the double-object construction, which requires a minimum value $M = 28$ to guarantee the extension of the *to*-dative construction. Conversely, there are $N = 35$ verbs in the *to*-dative construction, which requires a minimum value of $M = 26$ to justify the extension of the double-object construction. Therefore, the statistical analysis does not support a derivational analysis of two constructions.

To summarize, the sets of verbs attested in the dative constructions are semantically coherent to support the acquisition of the conditioning constraints. Furthermore, the Sufficiency Principle suggests that these constructions are productive, because a sufficiently large number of verbs that fit the semantic requirements are in fact attested in these syntactic forms. This accounts for the overgeneralization errors in child language, and further down the road, the extension of these constructions to novel but semantically appropriate verbs such as *email* and *text*.

But Baker's Paradox crucially involves further, and negative, constraints on the datives; namely, how do learners know that verbs such as *donate* do *not* participate in the double-object construction? The children must refine or retreat from the productive generalizations: eventually errors such as *I said her no* do disappear. We now confront Baker's Paradox in its full form.

6.3.3 Beyond the First Years of Life

Baker's Paradox just doesn't arise for the three-year-old.

Our five-million-word corpus represents virtually all of the child-directed English data in the public domain, but it does not represent the full range of complexity in the dative constructions. Very few Latinate verbs, which tend to resist the double-object construction and are at the heart of Baker's Paradox, can be found in child-directed English, and they do not appear at all in children's use of datives (Gropen et al. 1989). These vocabulary items are acquired relatively late during language acquisition (Mazurkewich and White 1984). We need a much higher volume of data for a full resolution of Baker's Paradox in language acquisition.

As noted earlier, the semantic conditioning for the double-object construction is very well supported in the child-directed corpus:

(36) caused possession $\implies$ double-object construction

But learners grow older, especially after exposure to more learned words in a school setting, their vocabulary of caused-possession verbs also increases. According to (36), these verbs are candidates for the double-object construction: some will be used in that form but others will not because they either cannot be (e.g., *dispatch*) or did not get a chance to (e.g., the *to*-dative construction is used instead). In order for (36) to remain productive, the numerical values of N and M — that is, the entire set of caused-possession verbs and the subset of which that appear in the double-object form — must satisfy the Sufficiency

6.3 Resolving Baker's Paradox

threshold. If this condition were to fail, or $(N - M) > \theta_N$, then caused possession will no longer automatically support the double-object construction, which results in retreating from the generalization in (36). Note that these considerations again naturally lead to the possibility of individual variation with respect to vocabulary as well as the productivity of the dative constructions.

So again, we are back in the numerical game. Unfortunately we can only speculate on the further development of the dative verbs, for currently there is no suitable corpus for an extended study of language acquisition. I will make use of several corpora to "bootstrap" a tentative estimate of an English speaker's dative verb vocabulary, on which the double-object rule (36) will be put to the test.

I first collected all the verbs from Levin 1993 that have the semantics of caused possession: these would be candidates for the double-object constructions. There are 115 caused-possession verbs that allow double-objects but 138 caused-possession verbs that do not, including the primarily Latinate class (e.g., *donate*), the "communication of propositions and propositional attitudes" class (e.g., *say*), and the "manner of speaking" class (e.g., *yell*). Taken at face value, the semantic condition of caused possession for the double-object construction cannot be maintained because there are many more exceptions than positive examples. No reasonable learning model should generalize a pattern that holds for only 45% of the data (115/253) to the entire set.

However, many of the verbs on Levin's lists are very rare, and their opportunities to participate in a dative construction (double-object or *to*-dative) would be even rarer. As noted throughout the current project, the learners' knowledge about virtually all aspects of language can only be established on the basis of a very small sample of the language; the data sparsity problem alone precludes us from having direct experience with all the verbs in Levin's classifications.

We thus need to "trim" Levin's verb lists to get a more realistic estimate of the dative experience of an English speaker. To do so I rely on the SUBTLEX-US Corpus (Brysbaert and New 2009), a 51-million-word spoken U.S. English corpus, as an anchoring point. The SUBTLEX-US Corpus provides both word frequency and parts of speech and has been found to provide the best correlates with a wide range of psycholinguistic measures involving the lexicon (Brysbaert and New 2009). My strategy is to use the frequency information of the verbs manually extracted from the child-directed corpus as a guideline, to obtain an approximate set of the dative verbs that English speakers are likely to encounter in their experience with the language.

I first identified the CHILDES dative verb with the lowest SUBTLEX-US frequency (*hoist*, 84 in 51 million). I then included all the verbs from Levin's exhaustive lists of 253 that (a) have a SUBTLEX-US frequency of at least 84 in 51 million and (b) are primarily used as a verb according to the part-of-speech information provided by the SUBTLEX-US Corpus. This maneuver produces 52 caused-possession verbs that can be expected to appear in a double-object form, and 40 caused-possession verbs that cannot. The two bootstrapped lists are provided below:[10]

(37) a. 52 caused-possession verbs expected to appear in double-object construction

ask, assign, bet, bring, carry, chuck, drag, extend, feed, flip, give, guarantee, haul, heave, hit, kick, lend, offer, owe, pass, pay, preach, promise, pull, push, quote, read, render, repay, save, sell, send, shoot, shove, show, sign, sing, slam, slap, slip, smuggle, sneak, spare, take, teach, tell, throw, toss, vote, wager, write, yield

b. 40 caused-possession verbs not expected to appear in double-object construction

administer, admit, announce, communicate, confess, contribute, convey, cry, declare, deliver, demonstrate, describe, dictate, dispatch, donate, explain, express, forfeit, holler, introduce, mention, propose, provide, recite, recommend, refer, repeat, restore, return, reveal, say, scream, shout, snap, squeal, submit, surrender, trust, whisper, yell

The verbs in (37a–b) are to interpreted as follows. Assuming that our bootstrapping method is reasonable, a "typical" English learner will have acquired 92 verbs of caused possession. I assume that all have had opportunities to be used in a dative construction: the 52 in (37a) will have appeared in the double-object construction, whereas the 40 in (37b) will not have. The rule (36), that caused possession enables the double-object construction, now fares better: its batting average goes from 0.45 (115/253) in the complete list of verbs to 0.58 (52/92) in the more restricted but also more plausible estimate of the

---

10 Again, the speakers I consulted disagreed with Levin's judgment to some extent, especially with respect to the verbs in (37b), because some speakers find verbs such as *explain*, *deliver*, and *provide* to be admissible in the double-object construction. Nevertheless, the discrepancies will not change the numerical relation between (37a) and (37b) in a qualitative way. As it stands, not nearly enough of the 92 verbs would appear in the double-object form to sufficiently support the productivity of the construction for the caused-possession verbs in English.

6.3 Resolving Baker's Paradox

verb vocabulary. Note further that in the SUBTLEX-US Corpus, the caused-possession verbs that allow double-objects are considerably more frequent that those that do not; on average, the verbs in (37a) appear almost three times as often as the verbs in (37b). This is consistent with our earlier results in section 6.3.2: the child-directed corpus, which contains higher-frequency verbs, strongly supports the double-object rule (36). If we break down the frequency of the 92 verbs into subsets ranked by frequency (Table 6.2), the productivity of the rule among higher frequency-words becomes very clear.

Table 6.2
Caused-possession verbs and their expected distribution with respect to the double-object construction

Top N	Yes	No	θ_N	Productive?
10	9	1	4	Yes
20	17	3	7	Yes
30	26	4	9	Yes
40	30	10	11	Yes
50	34	16	13	No
60	39	21	15	No
70	43	27	16	No
80	46	24	18	No
92	50	42	20	No

A young child, whose verb acquisition is strongly influenced by lexical frequency (Huttenlocher et al. 1991; Naigles and Hoff-Ginsberg 1998), will most certainly acquire an initial set of caused-possession verbs, of which an overwhelming majority appear in the double-object construction. For a speaker with a larger vocabulary of caused-possession verbs, however, the semantic condition for the double-object construction breaks down; the level of positive evidence does not even approach sufficiency. More specifically, out of $N = 92$ caused-possession verbs, only $M = 52$ can be expected to appear in the double-object construction: the learner will not have sufficient reason to generalize, which would require $92 - \theta_{92} = 72$.

What's next? The solution is very similar to those in the previous chapters, especially in the treatment of metrical stress in English and noun-plural formation in German. I assume that child learners are primed to discover productive generalizations within subdivisions of the data when no productive rule is supported over the entire dataset; lexicalization of the attested 52 verbs for the double-object construction must be a last resort. This, I suggest, completes the resolution to Baker's Paradox.

The phonological properties of words, which are closely linked with syntactic categories as made abundantly clear in the theory and acquisition of stress (section 4.2), provide a natural division of verbs. In fact, the phonology and syntax of the 92 verbs show a striking asymmetry:

(38) a. Of the 50 monosyllabic verbs, 42 allow the double-object construction and only 8 do not: cry, say, scream, shout, snap, squeal, trust, yell

b. Of the 42 polysyllabic verbs, 32 resist the double-object construction and only 10 allow it: assign, carry, extend, guarantee, offer, promise, render, repay, smuggle, wager

That is, for the set in (38a), an overwhelming majority of the 50, more than the Sufficiency threshold (at $50 - \theta_{50} = 38$), can be expected to appear in the double-object construction. By contrast, only a small fraction in the set (38b) will ever be used in the double-object construction. Thus, learners can conclude that monosyllabic verbs of caused possession allow double-objects but polysyllabic verbs do not, and the 10 that do (38b) are to be lexicalized.

Alternatively, learners may choose to focus on the stress properties of the verbs. Focusing on the stress patterns of the 42 polysyllabic verbs, the following observation can be made:

(39) a. Of the 12 initial-stress verbs, 6 allow the double-object construction (*carry, offer, promise, render, smuggle, wager*) while the other 6 do not (*demonstrate, donate, forfeit, holler, mention, whisper*).

b. Of the 30 verbs that do not have initial stress, only 4 allow the double-object construction (*assign, extend, guarantee, reply*).

It seems that noninitial stress, a well-known correlate with Latinate vocabulary, provides a very strong cue for the inadmissibility of the double-object construction.

Either way, the absence of an across-the-board productive rule force learners to partition the verbs into subsets, within which valid generalizations can be found. These generalizations include negative ones. A subset of verbs with,

6.3 Resolving Baker's Paradox

say, the right kind of semantics but wrong kind of phonology, does not automatically lead to the double-object construction because not enough members appear in the construction: the best the learners can do is to lexicalize those that do. The complement subset of verbs, with the appropriate phonological properties, can then be salvaged as a productive class that freely allows the double-object construction. The current approach to learning, then, accounts for the three essential facts about the dative constructions: (a) young children commit overgeneralization errors, (b) the phonological properties of the verb are correlated with the admissibility of the double-object condition that can be observed in school age children (Gropen et al. 1989) and adults (Mazurkewich and White 1984; Randall 1980), and (c) adult language retains productivity for a structurally well-defined set of verbs, which allow double-object extension when new verbs fitting the structural descriptions come into the language.

Finally, let's consider how children can eventually retreat from overgeneralization errors such as *I said her no*. Levin's classification contains a semantic subclass of communication verbs, where caused possession concerns the transmission of abstract information rather than the transfer of physical objects as in the more prototypical cases of *give*, *bring*, etc. I list those below:

(40) a. Verbs of transfer of a message ("verbs of type of communication messaged [differentiated by something like 'illocutionary force']"): ask, cite, pose, preach, quote, read, replay, show, teach, tell, write

b. Say-verbs ("verbs of communication of propositions and propositional attitudes"): admit, allege, announce, articulate, assert, communicate, confess, convey, declare, mention, propose, recount, repeat, report, reveal, say, state

Despite the very coherent semantic properties associated with the verbs in (40), only 11 in (40a) can be expected to appear in the double-object construction, whereas 17 in (40b) never will. On our SUBTLEX-US frequency-adjusted list, 8 of those in (40a) and 11 of those in (40b) are found. But for $N = 19$, $M = 8$ clearly falls below the Sufficiency threshold (at $19 - \theta_{19} = 13$). Thus, eventually English learners will recognize that verbs of communication, as a semantic class, do not allow the productive use of the double-object construction; the positively attested verbs in (40a) will be lexicalized. Since *say* is never used in the double-object construction by the caretakers, the children will eventually kick it off the list. This is similar to the development of *bring-brang*. For some children, the verb *bring* may fall under a transiently productive rule "ɪ → æ / __ ŋ" (section 4.1). When the rule loses its productivity as the vocabulary increases, children will only memorize the attested verbs (e.g., *ring*, *sing*) that follow the now lexicalized rule — forms such as *brang* will fade away.

To summarize our study of the dative constructions, the retreat from overgeneralizations follows the Tolerance Principle, with a twist. Recall that the emergence of the past-tense "*-d*" rule is made possible by an overwhelming number of regular verbs that trump the exceptions explicitly attested as non-*d* attaching: for the irregulars, then, the learner *knows* that they do not take -*d*. In the acquisition of the datives, however, generalization is possible only if a sufficiently large number of verbs are attested through positive evidence; the learner has no evidence one way or the other about the grammaticality of the unattested examples. In other words, I claim that for most English speakers, there is a difference between their rejection of *goed vs. **I donated charities money*. For *goed, the rejection comes from the knowledge that it is wrong because children hear *went*. But for **I donated charities money*, the rejection comes from uncertainty, or playing it safe. After all, if only a handful of exotic animals on the desert island are demonstrably tame, why should anyone take a chance?

****

I conclude with a methodological point on language acquisition. The availability of electronic databases in recent years has provided linguists with an ever expanding and readily accessible source of data. For instance, previous researchers (Boyd and Goldberg 2011; Bruening 2011a; Goldberg 2011) have made use of COCA (Davies 2008) as well as web examples in the analysis of *a*-adjectives. But the primary linguistic data for child language acquisition may be quite different from the type found in large-scale corpora. Negative evidence is surely absent in language acquisition, and positive evidence is pretty scant as well.

Indirect negative evidence, as I have argued, is to be avoided. It is too complex to be computationally feasible, and too coarse-grained to produce reliable learnability results in a realistic setting of language acquisition. This is a welcome result, because the absence of evidence *really* isn't the evidence of absence. If, as suggested by the Tolerance Principle, valid generalization requires a sufficient amount of positive evidence, then the absence of evidence will simply hold learners back to stay conservative, instead of following the

6.3 Resolving Baker's Paradox   213

path of logical fallacy. All the same, if a generalization does overshoot, learners can still backtrack to lexicalization if the amount of positive evidence drops below the Sufficiency threshold.

7 On Language Design

The numerical indulgence in the preceding pages, I hope, has not been a distraction. Before it can be considered a solution to Plato's and Darwin's Problem, the Tolerance Principle must meet the requirement of descriptive adequacy. I begin these concluding remarks by recapitulating the essential results of the present study. I then revisit the familiar theme of leaky grammars and discuss how the Tolerance Principle helps demarcate the boundary between the core and the periphery in a theory of language. Finally, I discuss the design features of language learning that may have been the best, and perhaps the only, path leading toward the rise of human language.

7.1 Computational Efficiency in Language Acquisition

Chomsky (2005, 6) envisions a new division of labor in the design of language, especially in light of factors not specific to language:

> The third factor falls into several subtypes: (a) principles of data analysis that might be used in language acquisition and other domains; (b) principles of structural architecture and developmental constraints that enter into canalization, organic form, and action over a wide range, including principles of efficient computation, which would be expected to be of particular significance for computational systems such as language. It is the second of these subcategories that should be of particular significance in determining the nature of attainable languages.

Executed to fruition, "third factors" will lead to a considerable simplification of the genetic endowment for language, providing a more plausible solution for Darwin's Problem, the evolution of language. Conceivably, the so-called Language Acquisition Device (Chomsky 1965) will no longer be necessary as a specialized, and independently evolved, module of the mind/brain. The variational model (Yang 2002) goes some way into this direction, because it employs general learning mechanisms (Bush and Mosteller 1951), "principles of data analysis that might be used in language acquisition and other domains." The development of the Tolerance Principle is more closely aligned with the principle of minimalist computation.

The current study can be viewed as a reverse engineering project: What are the computational mechanisms that help children discover rules and exceptions that permeate every corner of language? Chapter 2 reviews the design specifications. The input data to the learner, or indeed any linguistic corpus, shows a high degree of distributional sparsity, which poses significant difficulties for learning and challenges the conventional wisdom in the interpretation of linguistic performance. These statistical properties make the output of language acquisition even more impressive. Children are extraordinarily good at detecting rules and exceptions: rules are generalized and sometimes overgeneralized, but exceptions and other idiosyncrasies are almost always limited to a finite list of items.

The Tolerance Principle is developed as a theory of how children map the highly impoverished input to a highly sophisticated grammar. Methodologically, it assumes the psychological realism of linguistic machineries. Much of chapter 3 is devoted to establishing the Elsewhere Condition, a very traditional principle for handling rules and exceptions, as a performance model of real-time language processing. The Tolerance Principle, as a corollary of the Elsewhere Condition, provides an evaluation metric that favors faster grammars. Furthermore, allowing performance factors, which must involve non-linguistic components of cognition, to make decisions about the competence grammar eliminates the need for structural and/or learning-theoretic principles specific to language, including earlier proposals of the evaluation metric (e.g., symbol counting; Chomsky and Halle 1968).

Chapters 4 and 5 put the Tolerance Principle to the test, over a wide range of much studied problems in morphology and phonology. In chapter 4, I show that as an evaluation metric, the Tolerance Principle can guide learners toward the grammar of their language. Its recursive application is critical: the failure to identify productive patterns prompts learners to reassess the input data and modify their hypotheses. Of particular interest are the metrical stress of English and noun pluralization in German, which, when considered together, seem empirically irreconcilable. A rule that works for over 85% of words (initial stress in English) fails to generalize, yet a rule that covers less than 5% of words (the -*s* suffix in German) is completely productive. I show that the Tolerance Principle forces the data into well-partitioned structural classes, within which productive subregularities can be subsequently identified.

Chapter 5 looks at how productivity breaks down. When multiple alternations are available to a set of words, there may not be a sufficiently dominant winner. As a result, no productive rule will emerge and learners must resort to total lexicalization. Synchronically, lexicalization leads to gaps for words whose alternation is not attested in the language data; this is confirmed using lexical statistics in several well-known cases of ineffability. Diachronically, lexicalized words are vulnerable to leveling, because they succumb to productive rules, whose emergence can be similarly predicted on a numerical basis.

This approach offers a concrete program that integrates language acquisition and change as envisioned by Halle (1962). It has helped identify the likely causal mechanism that led to the recent changes in the Icelandic morphosyntactic system, and should be extended to additional cases of language change for which quantitative data is available.

Chapter 6 presents a learning strategy that generalizes from attested examples to the unobserved members of their shared class. Its most important application is to provide an alternative to indirect negative evidence, which I have argued to be computationally intractable and empirically ineffective. If children are more careful about when to generalize, the logical problem of retreating from overgeneralization is not as severe. According to the Sufficiency Principle, generalization only takes place after a sufficiently high amount of positive evidence has been accumulated. The inadmissibility of the *a*-adjectives in a pronominal position (*asleep cat*) is revealed by their distributional similarities with prepositional phrases. The conditions on the English datives can be constructed and accurately evaluated with relatively simple corpus data such child-directed speech, significantly reducing the structural prerequisites traditionally held to be innate features of Universal Grammar.

At this juncture, I can only speculate on the nature of the Elsewhere Condition, from which everything else follows. Although I am not aware of any analogy — never mind homology — of the Elsewhere Condition in the study of comparative cognition, the Gricean character of favoring specificity when possible can be observed elsewhere in language and possibly in even more general cognitive processes; see Aronoff and Lindsay 2015 for discussion.[1] To take a nonlinguistic example: the Fédération Internationale de Football Association (FIFA) has very clear regulations on the direct free kick (DFK) and the penalty kick (PK). According to the official *Rules of the Game*, a list of ten infringements results in a DFK, with a separate clause that "a penalty kick is awarded if any of the above ten offences is committed by a player inside his own penalty area." Nowhere does it specify that a more specific rule for the PK — foul in the penalty box — should trump the rule for the DFK (just the offense): it is simply taken for granted that specificity is privileged when

---

[1] I once lost a wager to Larry Horn for failing to coerce some undergraduate students into calling the thumb a finger: the "thumb" of course *is* a finger but the term is evidently preferred due to its specificity. It turns out that he had written on this very topic extensively (Horn 1989): Never bet against the house.

multiple rules are applicable. It is thus conceivable that the Elsewhere Condition itself is reducible to more general principles of human biology and cognition, which would remove yet another evolutionary innovation in the lineage of *Homo sapiens*. In the absence of comparative evidence, however, it is equally probable that the Elsewhere Condition is a derivative of the language faculty (e.g., Merge) that emerged in the uniquely human history of evolution. Finally, the Tolerance Principle, a solution to the problem of learning in the face of exceptions, may be applicable in nonlinguistic domains where the formal problem of learning is similar. The psychology of learning has produced a large body of literature on induction, generalization, and categorization, with some models developed specifically to deal with rules and exceptions that violate them (e.g., Anderson 1991; Medin and Schaffer 1978; Shepard 1987).[2] These will have to await future research.

7.2 Core and Periphery Revisited

The temptation for the wielder of the hammer is to see everything as a nail. Fortunately, we know enough about language acquisition to curb our enthusiasm. For all its utility in the inductive learning of language, it's wise not to stretch the Tolerance Principle to excess: we are not in for a revival of language-specific rules and constructions.

The initial goal of the current project was to account for the deficiencies in the variational model (Yang 2002), which was designed for the problem of parameter setting. Nothing here detracts from the validity of the variational model or the theoretical framework it assumes. The past fifty years of language acquisition research have produced, to my mind at least, indisputable evidence that the structure of linguistic forms is not arbitrary, and that innumerably many conceivable hypotheses are never entertained by either child or adult learners (Crain 1991; Smith and Tsimpli 1995; Tettamanti et al. 2004). These general constraints on language, whatever final form they turn out to take, appear universal and inviolable. They constitute prime examples for the classical argument from the poverty of the stimulus (Chomsky 1975), recently complemented with statistical and computational considerations (Berwick et al. 2011;

---

2 Although these models are designed to provide a statistical fit for experimental results with the valuation of model parameters. By contrast, the Tolerance Principle is parameter free, as it is the formal consequence of fundamental principles of language and cognition.

7.2 Core and Periphery Revisited

Legate and Yang 2002). For these, no form of learning is necessary, inductive or otherwise.

And I wish to remind readers that not all aspects of language acquisition follow the inductive generalization process in the present project. Drawing inspirations from biological evolution as well as mathematical theories of learning, the variational model captures the selectionist learning of language in a competition-based scheme: the nontarget but structurally possible grammars may be accessed prior to their ultimate demise. For instance, the null subject stage of child English, which lasts until the third birthday, bears the distributional as well as statistical marks of a topic-drop grammar such as Chinese (Yang 2002). For our purposes, it's not important how the range of grammatical options are theoretically characterized. It is not even important to call them "parameters." What *is* important, however, is that children spontaneously access and test this grammatical hypothesis without external stimuli. Language learning by selection, or unlearning as embodied in the variational model, appears indispensable for the explanation of child language. In such cases, no inductive learning, or the Tolerance Principle, will be needed.

There is another dimension in which the inductive learning of language-specific rules, the unlearning of innate parametric choices, and the non-learning of universal grammatical principles are differentiated. Consider the errors associated with inductive learning that we have studied in these pages (1a), in comparison to the ungrammatical examples in (1b):

(1)   a. *clinged
          *stridden
          *He donated the library some rare books.
    b. *car the (as opposed to "the car")
          *Is$_t$ the man that t tall is happy?
          *Whom$_t$ did you see John and t?

All the above forms are unacceptable to English speaker but that doesn't seem to be the whole story. The examples in (1a) fall into the realm of ambivalence or unfulfilled potential. As indicated clearly in our case studies, the ungrammaticality of the examples in (1a) is essentially accidental: there is no rhyme or reason why those specific lexical items fail to follow the rules that could have well applied to them. And it's often possible to find individuals or dialects in which the forms in (1a) are grammatical. By contrast, the forms in (1b) are uniformly repellent; it is highly unlikely that English speakers would find them acceptable under any circumstances.

I believe that the contrast between (1a) and (1b) stems from the core-vs.-periphery distinction (Chomsky 1981) that partly motivated the present project. The core and the periphery appear to elicit different reactions from language users, and their acquisition shows very distinct patterns. Chomsky's remarks were prophetic:

> Marked structures have to be learned on the basis of slender evidence too, so there should be further structure to the system outside of core grammar. We might expect that the structure of these further systems relates to the theory of core grammar by such devices as relaxing certain conditions of core grammar, processes of analogy in some sense to be made precise, and so on, though there will presumably be independent structure as well. (Chomsky 1981, 8)

> How do we delimit the domain of core grammar as distinct from marked periphery? In principle, one would hope that evidence from language acquisition would be useful with regard to determining the nature of the boundary or the propriety of the distinction in the first place, since it is predicted that the systems develop in quite different ways. (Chomsky 1981, 9)

One clear difference in the acquisition of the core vs. the periphery has to do with the nature of data that drives learners toward the target grammar. Putting aside the invariant and universal principles of language, which require no learning, we see that parameter setting—in the core—requires the unlearning of nontarget grammatical options, where the *token* frequency of disambiguating evidence determines the trajectory of learning. In the case of null subjects in child English, learners gradually move toward obligatory subject use from the cumulative effect of expletive subject sentences (e.g., *There is a cat in the bed*), which contradict the Chinese-type topic-drop grammar (Huang 1984; Hyams 1986; Yang 2002). Relying on a single instance in the sense of triggers does not lead to a robust or empirically adequate account of language learning. For the acquisition of the periphery, where inductive learning and the Tolerance Principle go hand in hand, the critical source of evidence comes from the *types* of (lexical) items. Hearing a rule exemplified by a single item, no matter how frequently, says nothing about its general applicability, which can only be established on the basis of sufficient diversity (N and θ_N).

The learning trajectories for the core and periphery are also markedly different. Parameter setting, as noted earlier, may show patterns of probabilistic

7.2 Core and Periphery Revisited

variation as children unlearn the unwanted options.[3] The inductive learning of the periphery, however, is typically characterized by discrete transitions. As the acquisition of English past-tense and dative constructions very clearly illustrates, children start off conservatively, not going beyond the specific lexical items that appear in the adult input. This stage of evidence accumulation is punctuated by the emergence of productive rules, which immediately gives rise to the potential for overgeneralization. Subsequently, overregularization errors may be ironed out over time, as in the inflection of irregular verbs. The rules themselves may be subsequently revised after additional input data is taken in, as in the learning of metrical stress and dative constructions.

Finally, and more speculatively, there seems to be a subtle though important difference in how input evidence contributes to the acquisition of the core and the periphery. Let's take the syntactic parameter of verb raising as an instance of the core grammar, and consider how a French-learning child may acquire the value of [+] as opposed to [−], the English option. Most of the input sentences in a French-speaking environment will not be informative. Examples such as (2a) provide no clear clue for the position of the verb and would have no effect on parameter selection. Only examples of the type in (2b), where the finite verb precedes "landmarks" such as adverb or negation, will push children toward the [+] option.

(2)   a. Jean voit Marie
          Jean sees Marie

      b. Jean voit$_t$ souvent/pas [t Marie]
          Jean sees often/not   Marie

More specifically, on encountering an example such as (2b), if learners select the [+] value of the parameter, the sentence will go through, the probability of the [+] value increases, and the learners get closer to the French grammar. If the [−] value has been selected, the sentence will fail to be analyzed and penalizes the [−] value which crucially, given the competition scheme in the

---

[3] It is a misinterpretation of the variational model that child language always shows the usage of competing variants (e.g., Snyder 2007). As has been made very clear in the mathematical formalism, the duration of variation — the time course required to drive out the nontarget competitors — depends on the frequency of disambiguating evidence in the input. Many parameters are acquired very early, and thus show that target usage form at the outset of children's language production, due to the high volume of critical data. See Legate and Yang 2002 and Yang 2012 for concise summaries.

variational model, also rewards the [+] value. In other words, positive evidence for a parametric value is negative evidence for the opposite value.

This is completely different from the acquisition of periphery rules such as the rules for a-adjectives and dative constructions (chapter 6). As I discussed at length on the basis of the child-directed English corpus, taking evidence for a hypothesis P as indirect negative evidence against $\neg P$ fails to produce the correct account of language learning. The fact that an adjective consistently, or even exclusively, appears in predicative use does not mean that it cannot be used attributively, and the fact that a verb never appears in a double-object construction does not mean that it cannot.

Plausibly, the difference between the core and the periphery reflects the fundamental distinction between functional and lexical categories in human language. The functional categories in the core such as Complementizer, Tense, v, etc., are innately provided, and their formal features are defined in opposition [±]. Parameter setting therefore takes place in a finite space of feature combinations, although variation and exceptions may still be present. For the periphery, however, the possible forms of rules — broadly speaking, including all of phonology and morphology (Bromberger and Halle 1989) — are much more arbitrary, and acquisition studies suggest that even "unnatural" rules do not seem to cause special difficulty for young language learners (Seidl and Buckley 2005).[4] As shown in the numerous case studies presented in the preceding pages, it is clear that no amount of innate knowledge can replace the need for data-driven induction and lexical learning. If anything is learned in language, it's words.

Table 7.1 provides a tentative summary of the key differences between the grammatical core and periphery from the perspective of language acquisition.

Returning to the controversy over the core-vs.-periphery distinction, and the much longer debate on the treatment of exceptions (Chapter 1), it is impossible to know, a priori, which compartment of the grammar a specific phenomenon belongs to, and considerable crosslinguistic research will always be necessary. But I hope to have shown that child language, and the computational mechanisms of language acquisition, have much to offer to the theory of grammar. A principled division between two kinds of linguistic knowledge is not only possible; it in fact can be drawn.

---

4 Although they would still be expected to follow the general principles of language. It's hard to imagine that Structure Dependence shuts off in a peripheral construction where a movement operation targets, say, the third word in a string.

Table 7.1
The core and the periphery in comparison

	Core	Periphery
Example	Null subjects	Dative construction
Ontogeny	Internally controlled	Externally derived
Evidence	Tokens	Types
Distribution	Universal	Language particular
Development	Continuous variation	Discrete productivity
Grammatical category	Functional	Lexical
Learning model	Selection	Induction

7.3 The Ecology of Language Learning

Nature does not build things from scratch. Evolution is a tinkerer: it recycles and reuses old parts. When a new piece of equipment arrives, it has to work together with existing structures. Language is the latest step in the rise of *Homo sapiens* but it must have been built upon other components of the mind that existed before language, and would have to function within the ecological constraints of the human condition.

A nonnegotiable feature of language is learnability. The native language must be acquirable from a relatively simple, and sparsely distributed, sample of primary linguistic data, and with biologically plausible mechanisms of learning that undergo the normal course of growth and development. Without taking the developmental constraints into account, we may create artificial problems that would never arise in nature. For instance, Baker's Paradox, which is often presented with examples using *give* and *donate*, is almost surely self-inflicted by linguists and for linguists. When children are learning *give* and its syntactic properties, *donate* almost surely will not even enter into consideration, for its acquisition is probably still a few years away. And even if *give* and *donate*, *tell* and *say*, etc. are presented all at the same time, who in their right mind would generalize from a handful of items to an open-ended class? By the same token — this is a bit uncharitable but no less true — when linguists discover an apparent pattern in the data through a statistical analysis of the corpus or just through a good old-fashioned hunch, it does not automatically gain empirical status unless accompanied by evidence of its cognitive accessibility to the human mind. To take an extreme example; about 37% of modern English words are nouns, but it is difficult to see what if anything this numerical curiosity can tell us about language and the mind.

The statistical reality of language provides another important, and interconnected, constraint on language learning and language design. The sparsity of linguistic data has been rightly vilified in natural language processing (Jelinek 1998). In simplistic models of language, the number of parameters that need valuation quickly overwhelms the corpora. Two conclusions are immediate when the learning problem is placed in an ecological setting of language acquisition. First, and as a matter of logic, the child's language model must be sufficiently abstract, capable of covering a wide range of data that may seem superficially different, as well as sufficiently simple, such that the learning decisions can be made correctly with a relatively small amount of data. Second, and this is less obvious but has been implicit all along, language learning not only needs to overcome impoverished data but is also *enabled* by impoverished data. Let me unpack this rather paradoxical argument.

An often-overlooked feature of language acquisition is the striking uniformity of the grammar attained. This is remarkable because as a biological trait, there is bound to be individual-level variation in children's ability to analyze and assimilate linguistic data, an issue I have devoted some attention to in section 4.1.2 in discussing the acquisition of inflectional morphology by Adam, Abe, and Eve; see also Legate and Yang 2007. Moreover, considerable variation, in quantity as well as quality, has been observed in the linguistic input that children receive, with socioeconomic factors a primary determinant (Hart and Risley 1995). There is no doubt that some of this individual variation will persist into adulthood (Sankoff and Blondeau 2007), which only makes more striking "the uniformity of abstract patterns of variation which are invariant in respect to particular levels of usage" (Labov 1972b, 120), as well as the uniformity in the direction and outcome of linguistic change (Labov 2007; Labov et al. 2006). The uniformity of language acquisition is sharply contrasted with current natural language processing systems, which are heavily dependent on the genre and size of the training corpus and do not adapt easily to other domains (Gildea 2001; McClosky, Charniak, and Johnson 2010).

The uniformity of language acquisition suggests a species-general biological capacity for language (Chomsky 1965). At the same time, the computational mechanism of language acquisition must be highly robust and resilient, yielding largely invariant grammars despite the variation in the input and across individuals. When the input data is sparsely distributed, uniformity can only be achieved on the basis of relatively high-frequency items available to all

7.3 The Ecology of Language Learning

learners: in other words, Zipf's long tail, where the majority of linguistic units (e.g., words) reside, is essentially omissible noise.

In fact, most words *must* be omissible in order for language acquisition to succeed. As I have stressed throughout, children acquire language incrementally, seeking productive generalizations along the way. Had a larger vocabulary been used, the stress rules of English and the antiattributive properties of *a*-adjectives would not have been learnable. The price of productivity is steep — recall the small growth of the Tolerance threshold θ_N — and gets steeper the longer children have to wait.

A similar situation arises in the acquisition of categorical rules, which also appear to be underrepresented in the data. In earlier work, I investigated the status of grammar in early child English (Yang 2013a). A fully productive rule "NP $\rightarrow$ D N" suggests that the determiner (D; *a/n* and *the*) can be interchangeably used with singular nouns (N). In numerous corpus analyses, both children and adults produce fairly low values of combinatorial diversity: typically only 20% to 40% of nouns that appear with either determiner are paired with both (Pine and Lieven 1997; Valian, Solt, and Stewart 2009), giving rise to the impression that young children do not have abstract rules but but rely on memorizing lexically specific combinations from adult input (Tomasello 2003). Yet a rigorous statistical test (Yang 2013a) shows that even very young children produce the level of diversity that, while low, is expected under a categorical rule that independently combines determiners and nouns.

Let's consider a concrete case that involves, again, Adam. In his speech transcripts, Adam produced 3,729 determiner-noun combinations with 780 distinct nouns. Of these, only 32.2% appeared with both determiners, which is similar to the expected value of 33.7% under the abstract NP rule; see Yang 2013a for details. Adam's mother, who was recorded in the same corpus, produced a diversity measure of 30.3% out of 914 nouns. Even among the 469 nouns used at least twice, which provided opportunities to be used with both determiners, only just over half (260) did so. To learn the NP rule, then, Adam must, and apparently did at a very young age, generalize from a small subset of nouns with attested interchangeable determiners to all nouns.

What could account for this massive leap of faith? Only if children attend to a small amount of data. The developmental literature offers the idea of "less is more" (Elman 1993; Newport 1990): the maturational constraints place a limit on the processing capacity of young children, which may turn out to be beneficial for language acquisition. Under the sparsity of language distribution, the acquisition of the determiner rule (and by extension, any rules) would be impossible if the learner required evidence from all or even most of the participating units. But if the children can only retain and learn from the most frequent items, the odds of acquiring productive rules improve considerably. Furthermore, under the machinery of the Tolerance Principle, it is also easier for rules that operate over a small class of items to clear the productivity threshold (Table 3.1).

Consider again the determiner-noun combinations produced by Adam's mother, for whom only 277 out of the 914 nouns are paired with both *a/n* and *the*. The Principle of Sufficiency transparently fails to detect interchangeability as a productive feature of the DP rule since the threshold is 770 ($\theta_{914} = 134$). But if Adam were only to learn from the 50 most frequent nouns, he would notice that almost all of them — 43 to be precise — are paired with both determiners. On this much smaller subset of data where $N = 50$, there *is* sufficient evidence for generalization: the 7 nouns that appear exclusively with only one determiner are below the tolerance threshold $\theta_{50} = 12$. For the $N = 100$ most frequent nouns, 83 are paired with both determiners: the 17 loners are again below the tolerance threshold $\theta_{100} = 22$. The vocabulary size of children at the age when they show productivity of the NP rule cannot exceed a few hundred words (Fenson et al. 1994). They must have acquired the rule on a very small set of high frequency nouns, almost all of which will show interchangeability with both determiners; the rest is just noise.

So the sparsity of language is much more of a blessing than a curse, an unusual conclusion that we hope to explore and exploit in future studies of language and its engineering applications (Chan 2008; Lignos 2010; Zhao and Marcus 2009). Indeed, it appears to go hand in hand with the developmental constraints on language and cognition. A sledgehammer like the Tolerance Principle may be the best hope for young children to pick up the rules of language before learning to climb trees or tie their shoes, thereby ensuring the successful transmission of our unique biological and cultural heritage.

*

For outsiders like me, generative grammar was appealing because it was familiar yet different. The first impression was the overwhelming richness and complexity found in the world's languages, which quickly put to rest seductively simple but ultimately simple-minded ideas (insert the latest buzzword from the Science section of the *New York Times*). Furthermore, the axiomatic and deductive nature of linguistics marks a clean break from the traditional methods in the social and behavioral sciences, which continue to loop through the cycle of data collection, statistical analysis, and repeat. In the best kind of linguistic practice, simple hypotheses can be formulated precisely such that their empirical consequences of nontrivial depth can be worked out by mechanical means. Theoretical developments take place well before the collection and verification of data: this element of suspense was comfortably familiar to those of us who write programs for a living. Occasionally, we do come across general principles of language that connect a wide range of phenomena; no need to bake each separately into the theory, or to invoke yet another variable in the model of regression.

As I hope to have made clear in these pages, language and language acquisition also contain a strikingly mechanical element. For all the unpredictabilities of young children, evident even though I have only direct experience with two, there are trends and regularities that all learners follow before converging on a largely invariant grammar shared across the community. Empirical findings in child language, reinforced by evolutionary considerations, suggest that the mechanisms of language acquisition must be simple: perhaps no more than keeping track of two numbers.

In my view, the most important conclusion from the present study is not whether the Tolerance Principle is ultimately correct. It is much more important that something like the Tolerance Principle can be established in the first place; by working out the axioms of language and cognition to their deductive ends — which is why it had to be in the form of an equation. This still strikes me as the most exciting aspect of generative grammar, even if the solution on offer turned out to be a lucky guess.

Bibliography

Akhtar, N. and Tomasello, M. (1997). Young children's productivity with word order and verb morphology. *Developmental Psychology*, 33(6):952–965.

Albright, A. (2002). Islands of reliability for regular morphology: Evidence from Italian. *Language*, 78(4):684–709.

Albright, A. (2003). A quantitative study of Spanish paradigm gaps. In *Proceedings of the 22nd West Coast Conference on Formal Linguistics*, pages 1–14. Cascadilla, Somerville, MA.

Albright, A. (2005). The morphological basis of paradigm leveling. In Downing, L. J., Hall, T. A., and Raffelsiefen, R., editors, *Paradigms in phonological theory*, pages 17–43. Oxford University Press, Oxford.

Albright, A. (2006). Cautious generalization of inflectional morphology, and its role in defectivity. Paper presented at the conference on defective paradigms.

Albright, A. (2009). Lexical and morphological conditioning of paradigm gaps. In Rice, C. and Blaho, S., editors, *Modeling ungrammaticality in Optimality Theory*, pages 117–164. Equinox, London.

Albright, A., Andrade, A., and Hayes, B. (2001). Segmental environments of Spanish diphthongization. In Albright, A. and Cho, T., editors, *UCLA Working Papers in Linguistics 7: Papers in Phonology 5*, pages 117–151. Cascadilla, Somerville, MA.

Albright, A. and Hayes, B. (2003). Rules vs. analogy in English past tenses: A computational/experimental study. *Cognition*, 90(2):119–161.

Alegre, M. and Gordon, P. (1999). Frequency effects and the representational status of regular inflections. *Journal of Memory and Language*, 40(1):41–61.

Allen, M. and Badecker, W. (2002). Inflectional regularity: Probing the nature of lexical representation in a cross-modal priming task. *Journal of Memory and Language*, 46(4):705–722.

Allen, S. (1996). *Aspects of argument structure acquisition in Inuktitut*. John Benjamins Publishing, Amsterdam.

Ambridge, B., Kidd, E., Rowland, C. F., and Theakston, A. L. (2015). The ubiquity of frequency effects in first language acquisition. *Journal of Child Language*, 42(02):239–273.

Anderson, J. R. (1991). The adaptive nature of human categorization. *Psychological Review*, 98(3):409.

Anderson, S. R. (1969). *West Scandinavian vowel systems and the ordering of phonological rules*. PhD thesis, MIT.

Anderson, S. R. (1974). *The organization of phonology*. Academic Press, New York.

Anderson, S. R. (1988a). Morphological change. In Newmeyer, F. J., editor, *Linguistics: The Cambridge survey*, pages 146–191. Cambridge University Press, Cambridge.

Anderson, S. R. (1988b). Morphology as a parsing problem. *Linguistics*, 26(4):521–544.

Anderson, S. R. (1992). *A-morphous morphology*. Cambridge University Press, Cambridge.

Anderson, S. R. (2010). Failing one's obligations: Defectiveness in Rumantsch reflexes of DĒBĒRE. In Baerman, M., Corbett, G. G., and Brown, D., editors, *Defective paradigms: Missing forms and what they tell us*, pages 19–34. Oxford University Press, Oxford.

Anderson, S. R. (2015). Morphological change. In Bowern, C. and Evans, B., editors, *The Routledge Handbook of Historical Linguistics*, pages 264–285. Routledge, New York.

Anderwald, L. (2009). *The morphology of English dialects: Verb-formation in non-standard English*. Cambridge University Press, Cambridge.

Anderwald, L. (2013). Natural language change or prescriptive influence?: Throve, dove, pled, drug and snuck in 19th-century American English. *English World-Wide*, 34(2):146–176.

Anglin, J. M. (1993). Vocabulary development: A morphological analysis, with commentary by George A. Miller and Pamela C. Wakefield. *Monographs of the Society for Research in Child Development*, pages i–186.

Angluin, D. (1980). Inductive inference of formal languages from positive data. *Information and Control*, 45(2):117–135.

Anshen, F. and Aronoff, M. (1988). Producing morphologically complex words. *Linguistics*, 26(4):641–656.

Anttila, A. (2002). Morphologically conditioned phonological alternations. *Natural Language and Linguistic Theory*, 20(1):1–42.

Armstrong, S. L., Gleitman, L. R., and Gleitman, H. (1983). What some concepts might not be. *Cognition*, 13(3):263–308.

Aronoff, M. (1976). *Word formation in generative grammar*. MIT Press, Cambridge, MA.

Aronoff, M. and Lindsay, M. (2015). Partial organization in languages: La langue est un système où la plupart se tient. *Proceedings of the 8th Décembrettes*, pages 1–14.

Atserias, J., Casas, B., Comelles, E., González, M., Padró, L., and Padró, M. (2006). Freeling 1.3: Syntactic and semantic services in an open-source nlp library. In *Proceedings of LREC*, volume 6, pages 48–55.

Baayen, R. H. (1989). *A corpus-based approach to morphological productivity: Statistical analysis and psycholinguistic interpretation*. PhD thesis, Vrije Universiteit Amsterdam.

Baayen, R. H. (2009). Corpus linguistics in morphology: Morphological productivity. In Ludeling, A. and Kyto, M., editors, *Corpus linguistics. An international handbook*, pages 900–919. Mouton de Gruyter.

Baayen, R. H., Dijkstra, T., and Schreuder, R. (1997). Singulars and plurals in Dutch: Evidence for a parallel dual-route model. *Journal of Memory and Language*, 37(1):94–117.

Baayen, R. H. and Lieber, R. (1991). Productivity and English derivation: A corpus-based study. *Linguistics*, 29(4):801–843.

Baayen, R. H., McQueen, J. M., Dijkstra, T., and Schreuder, R. (2003). Frequency effects in regular inflectional morphology: Revisiting Dutch plurals. In *Morphological structure in language processing*, pages 355–390. Mouton, Berlin.

Baayen, R. H., Piepenbrock, R., and Gulikers, L. (1996). CELEX2. Linguistic Data Consortium: LDC96L14.

Baayen, R. H. and Renouf, A. (1996). Chronicling the Times: Productive lexical innovations in an English newspaper. *Language*, 72(1):69–96.

Baerman, M. (2008). Historical observations on defectiveness: The first singular non-past. *Russian Linguistics*, 32(1):81–97.

Baerman, M. and Corbett, G. G. (2010). Defectiveness: Typology and diachrony. In Baerman, M., Corbett, G. G., and Brown, D., editors, *Defective paradigms: Missing forms and what they tell us*, pages 19–34. Oxford University Press, Oxford.

Baerman, M., Corbett, G. G., and Brown, D., editors (2010). *Defective paradigms: Missing forms and what they tell us*. Oxford University Press, Oxford.

Baker, C. L. (1979). Syntactic theory and the projection problem. *Linguistic Inquiry*, 10(4):533–581.

Baker, M. (1988). *Incorporation: A theory of grammatical function changing*. University of Chicago Press, Chicago.

Baker, M. (2001). *The atoms of language: The mind's hidden rules of grammar*. Basic Books, New York.

Baker, R. and Smith, P. (1976). A psycholinguistic study of English stress assignment rules. *Language and Speech*, 19(1):9–27.

Bakovic, E. (2013). *Blocking and complementarity in phonological theory.* Equinox, London.

Balota, D. A. and Duchek, J. M. (1988). Age-related differences in lexical access, spreading activation, and simple pronunciation. *Psychology and Aging*, 3(1):84.

Balota, D. A., Yap, M. J., Cortese, M. J., Hutchison, K. A., Kessler, B., Loftis, B., Neely, J. H., Nelson, D. L., Simpson, G. B., and Treiman, R. (2007). The English Lexicon Project. *Behavior Research Methods*, 39(3):445–459.

Baptista, B. O. (1984). English stress rules and native speakers. *Language and speech*, 27(3):217–233.

Baroni, M. (2009). Distributions in text. In Lüdeling, A. and Kyöto, M., editors, *Corpus linguistics: An international handbook*, pages 803–821. Mouton de Gruyter, Berlin.

Baroni, M. and Lenci, A. (2010). Distributional memory: A general framework for corpus-based semantics. *Computational Linguistics*, 36(4):673–721.

Baronian, L. (2005). *North of phonology*. PhD thesis, Stanford University.

Baronian, L. and Kulinich, E. (2012). Paradigm gaps in whole word morphology. *Irregularity in morphology (and beyond). Studia typologica*, 11:81–100.

Bartke, S., Rösner, F., Streb, J., and Wiese, R. (2005). An ERP-study of German 'irregular' morphology. *Journal of Neurolinguistics*, 18(1):29–55.

Bauer, L. (1983). *English word-formation*. Cambridge Textbooks in Linguistics. Cambridge University Press, Cambridge.

Bauer, L. (2001). *Morphological productivity*. Cambridge University Press, Cambridge.

Beard, R. (1995). *Lexeme-morpheme base morphology*. SUNY Press, Albany.

Beck, S. and Johnson, K. (2004). Double objects again. *Linguistic Inquiry*, 35(1):97–124.

Becker, M., Ketrez, N., and Nevins, A. (2011). The surfeit of the stimulus: Analytic biases filter lexical statistics in Turkish laryngeal alternations. *Language*, 87(1):84–125.

Bellugi, U. and Brown, R. (1964). *The acquisition of language: Report of the Fourth Conference sponsored by the Committee on Intellective Processes Research of the Social Science Research Council*. Monographs of the Society for Research in Child Development Volume 29. Wiley, New York.

Berko, J. (1958). The child's learning of English morphology. *Word*, 14(2–3):150–177.

Berwick, R. (1985). *The acquisition of syntactic knowledge*. MIT Press, Cambridge, MA.

Berwick, R. C., Pietroski, P., Yankama, B., and Chomsky, N. (2011). Poverty of the stimulus revisited. *Cognitive Science*, 35(7):1207–1242.

Biberauer, T., Holmberg, A., Roberts, I., and Sheehan, M., editors (2010). *Parametric variation: Null subjects in minimalist theory*. Cambridge University Press, Cambridge.

Bickerton, D. (1995). *Language and human behavior*. University of Washington Press, Seattle, WA.

Bierwisch, M. (1967). Syntactic features in morphology: General problems of so-called pronominal inflection in German. In *To honor Roman Jakobson: Essays on the occasion of his 70th birthday*, volume 1, pages 239–270. Mouton, The Hague.

Bikel, D. M. (2004). Intricacies of Collins' parsing model. *Computational Linguistics*, 30(4):479–511.

Bittner, D. (2000a). Gender classification and the inflectional system of German nouns. *Trends in Linguistics Studies and Monographs*, 124:1–24.

Bittner, D. (2000b). Sprachwandel durch Spracherwerb? Pluralerwerb. In Bittner, A., Bittner, D., and Köpcke, K.-M., editors, *Systemorganisation in Phonologie, Morphologie und Syntax*, pages 123–140. Olms, Hildesheim.

Bloch, B. (1947). English verb inflection. *Language*, 23(4):399–418.

Bloom, L. (1970). *Language development: Form and function in emerging grammar.* MIT Press, Cambridge, MA.

Bloom, P. (1990). Subjectless sentences in child language. *Linguistic Inquiry*, pages 491–504.

BNC Consortium (1995). British National Corpus. University of Oxford.

Bobrow, S. A. and Bell, S. M. (1973). On catching on to idiomatic expressions. *Memory & Cognition*, 1(3):343–346.

Bolhuis, J. J., Tattersall, I., Chomsky, N., and Berwick, R. C. (2014). How could language have evolved? *PLoS Biol*, 12(8):e1001934.

Bolinger, D. L. M. (1971). *The phrasal verb in English*. Harvard University Press, Cambridge, MA.

Bolozky, S. (1999). *Measuring productivity in word formation: The case of Israeli Hebrew*, volume 27. Brill.

Bortfeld, H., Morgan, J. L., Golinkoff, R. M., and Rathbun, K. (2005). Mommy and me: Familiar names help launch babies into speech-stream segmentation. *Psychological Science*, 16(4):298–304.

Botha, R. P. (1969). *The function of the lexicon in transformational generative grammar*. Mouton, Berlin.

Bouldin, J. M. (1990). *The syntax and semantics of postnominal adjectives in English*. PhD thesis, University of Minnesota.

Bowerman, M. (1973). *Early syntactic development: A cross-linguistic study with special reference to Finnish*, volume 11. Cambridge University Press, Cambridge.

Bowerman, M. (1982). Reorganizational process in lexical and syntactic development. In Wanner, E. and Gleitman, L. R., editors, *Language acquisition: The state of the art*, pages 319–346. Cambridge University Press, New York.

Bowerman, M. (1988). The 'no negative evidence' problem: How do children avoid constructing an overly general grammar? In Hawkins, J. A., editor, *Explaining language universals*, pages 73–101. Basil Blackwell, Oxford.

Bowerman, M. and Croft, W. (2008). The acquisition of the English causative alternation. In Bowerman, M. and Brown, P., editors, *Crosslinguistic perspectives on argument structure: Implications for learnability*, pages 279–307. Erlbaum.

Boyd, J. K. and Goldberg, A. E. (2011). Learning what not to say: The role of statistical preemption and categorization in a-adjective production. *Language*, 87(1):55–83.

Boyé, G. and Cabredo Hofherr, P. (2010). Defectivity as stem suppletion in French and Spanish verbs. In Baerman, M., Corbett, G. G., and Brown, D., editors, *Defective paradigms: Missing forms and what they tell us*, pages 35–52. Oxford University Press, Oxford.

Braine, M. D. (1963). The ontogeny of English phrase structure: The first phase. *Language*, 39(1):1–13.

Braine, M. D. (1971). On two types of models of the internalization of grammars. In Slobin, D. I., editor, *The ontogenesis of grammar: A theoretical symposium*, pages 153–186. Academic Press, New York.

Brame, M. K. and Bordelois, I. (1973). Vocalic alternations in Spanish. *Linguistic Inquiry*, 4(2):111–168.

Brants, S., Dipper, S., Eisenberg, P., Hansen-Schirra, S., König, E., Lezius, W., Rohrer, C., Smith, G., and Uszkoreit, H. (2004). TIGER: Linguistic interpretation of a German corpus. *Research on Language and Computation*, 2(4):597–620.

Bresnan, J. and Ford, M. (2010). Predicting syntax: Processing dative constructions in American and Australian varieties of English. *Language*, 86(1):168–213.

Bresnan, J. and Nikitina, T. (2009). On the gradience of the dative alternation. In Wee, L. and Uyechi, L., editors, *Reality explorations and discovery: Pattern interaction in language and life*, pages 161–184. CSLI Publications, Stanford, CA.

Brill, E. (1995). Transformation-based error-driven learning and natural language processing: A case study in part of speech tagging. *Computational Linguistics*, 21(4):543–565.

Bromberger, S. and Halle, M. (1989). Why phonology is different. *Linguistic Inquiry*, 20(1):51–70.

Brooks, P. J. and Tomasello, M. (1999). How children constrain their argument structure constructions. *Language*, 75(4):720–738.

Brown, D. and Hippisley, A. (2012). *Network morphology: A defaults-based theory of word structure*, volume 133. Cambridge University Press, Cambridge.

Brown, R. (1973). *A first language: The early stages*. Harvard University Press, Cambridge, MA.

Brown, R. and Bellugi, U. (1964). Three processes in the child's acquisition of syntax. *Harvard Educational Review*, 34(2):133–151.

Brown, R. and Fraser, C. (1963). The acqisition of syntax. In Cofer, C. and Musgrave, B., editors, *Verbal behavior and learning: Problems and processes*, pages 158–201. McGraw-Hill, New York.

Brown, R. and Hanlon, C. (1970). Derivational complexity and the order of acquisition in child speech. In Hayes, J. R., editor, *Cognition and the development of language*, pages 11–53. Wiley, New York.

Brown, R. W. (1957). Linguistic determinism and the part of speech. *The Journal of Abnormal and Social Psychology*, 55(1):1–5.

Bruening, B. (2010). Double object constructions disguised as prepositional datives. *Linguistic Inquiry*, 41(2):287–305.

Bruening, B. (2011a). A-adjectives again. http://lingcomm.blogspot.com/.

Bruening, B. (2011b). A-adjectives are PPs, not adjectives. http://lingcomm.blogspot.com/.

Brysbaert, M. and New, B. (2009). Moving beyond Kučera and Francis: A critical evaluation of current word frequency norms and the introduction of a new and improved word frequency measure for American English. *Behavior Research Methods*, 41(4):977–990.

Bush, R. R. and Mosteller, F. (1951). A mathematical model for simple learning. *Psychological Review*, 68(3):313–323.

Butt, J. and Benjamin, C. (1988). *A new reference grammar of modern Spanish*. Edward Arnold, London.

Bybee, J. (1985). *Morphology: A study of the relation between meaning and form*. John Benjamins, Philadelphia.

Bybee, J. and Moder, C. L. (1983). Morphological classes as natural categories. *Language*, 59(2):251–270.

Bybee, J. and Pardo, E. (1981). Morphological and lexical conditioning of rules: Experimental evidence from Spanish. *Linguistics*, 19:937–968.

Bybee, J. L. (1995). Regular morphology and the lexicon. *Language and Cognitive Processes*, 10(5):425–455.

Bybee, J. L. and Slobin, D. (1982). Rules and schemas in the development and use of the English past tense. *Language*, 58(2):265–289.

Cacciari, C. and Tabossi, P. (1988). The comprehension of idioms. *Journal of Memory and Language*, 27(6):668–683.

Campbell, L. (2004). *Historical linguistics: An introduction*. MIT Press, Cambridge, MA, 2nd edition.

Caprin, C. and Guasti, M. T. (2009). The acquisition of morphosyntax in Italian: A cross-sectional study. *Applied Psycholinguistics*, 30(1):23–52.

Caramazza, A. (1997). How many levels of processing are there in lexical access? *Cognitive Neuropsychology*, 14(1):177–208.

Caramazza, A., Laundanna, A., and Romani, C. (1988). Lexical access and inflectional morphology. *Cognition*, 28(3):297–332.

Carey, S. and Bartlett, E. (1978). Acquire a single new word. *Child Language Development*, 15.

Cerella, J. and Fozard, J. L. (1984). Lexical access and age. *Developmental Psychology*, 20(2):235.

Chan, E. (2008). *Structures and distributions in morphology learning*. PhD thesis, University of Pennsylvania.

Charniak, E. and Johnson, M. (2005). Coarse-to-fine n-best parsing and maxent discriminative reranking. In *Proceedings of the 43rd Annual Meeting on Association for Computational Linguistics*, pages 173–180. Association for Computational Linguistics.

Chater, N. and Vitányi, P. (2007). Ideal learning of natural language: Positive results about learning from positive evidence. *Journal of Mathematical Psychology*, 51(3):135–163.

Chomsky, N. (1951). Morphophonemics of Modern Hebrew. Master's thesis, University of Pennsylvania. Published by Garland, New York, 1979.

Chomsky, N. (1955). The logical structure of linguistic theory. Ms., Harvard University and MIT. Revised version published by Plenum, New York, 1975.

Chomsky, N. (1958). [Review of Belevitch 1956]. *Language*, 34(1):99–105.

Chomsky, N. (1959). A review of B.F. Skinner's *Verbal Behavior*. *Language*, 35(1):26–58.

Chomsky, N. (1965). *Aspects of the theory of syntax*. MIT Press, Cambridge, MA.

Chomsky, N. (1970). Remarks on nominalization. In Jacobs, R. A. and Rosenbaum, P., editors, *Readings in English transformational grammar*, pages 184–221. Ginn, Waltham, MA.

Chomsky, N. (1975). *Reflections on language*. Pantheon, New York.

Chomsky, N. (1981). *Lectures in government and binding*. Foris, Dordrecht.

Chomsky, N. (1986). *Knowledge of language: Its nature, origins, and use*. Praeger, New York.

Chomsky, N. (1995). *The minimalist program*. MIT Press, Cambridge, MA.

Chomsky, N. (2001). Derivation by phase. In Kenstowicz, M., editor, *Ken Hale: A life in language*, pages 1–52. MIT Press, Cambridge, MA.

Chomsky, N. (2005). Three factors in language design. *Linguistic Inquiry*, 36(1):1–22.

Chomsky, N. and Halle, M. (1968). *The sound pattern of English*. MIT Press, Cambridge, MA.

Chung, S. (1998). *The design of agreement: Evidence from Chamorro*. University of Chicago Press, Chicago.

Cinque, G. (1994). On the evidence for partial N-movement in the Romance DP. In Koster, J., Pollock, J.-Y., Rizzi, L., and Zanuttini, R., editors, *Paths toward Universal Grammar*, pages 85–110. Georgetown University Press, Georgetown.

Cinque, G. (2010). *The syntax of adjectives: a comparative study*. MIT Press, Cambridge, MA.

Clahsen, H. (1997). The representation of participles in the German mental lexicon: Evidence for the dual-mechanism model. *Yearbook of Morphology*, pages 73–95.

Clahsen, H. (1999). Lexical entries and rules of languge: A multidisciplinary study of German inflection. *Behavioral and Brain Sciences*, 22:991–1069.

Clahsen, H., Aveledo, F., and Roca, I. (2002). The development of regular and irregular verb inflection in Spanish child language. *Journal of Child Language*, 29(3):591–622.

Clahsen, H., Eisenbeiss, S., and Sonnenstuhl, I. (1997). Morphological structure and the processing of inflected words. *Theoretical Linguistics*, 23(3):201–249.

Clahsen, H., Hadler, M., and Weyerts, H. (2004). Speeded production of inflected words in children and adults. *Journal of Child Language*, 31(3):683–712.

Clahsen, H. and Penke, M. (1992). The acquisition of agreement morphology and its syntactic consequences: New evidence on German child language from the Simone Corpus. In Meisel, J., editor, *The acquisition of verb placement*, pages 181–234. Kluwer, Dordrecht.

Clahsen, H. and Rothweiler, M. (1993). Inflectional rules in children's grammars: Evidence from German participles. *Yearbook of Morphology*, pages 1–34.

Clahsen, H., Rothweiler, M., Woest, A., and Marcus, G. (1992). Regular and irregular inflection in the acquisition of German noun plurals. *Cognition*, 45:225–255.

Clark, E. V. (1987). The principle of contrast: A constraint on language acquisition. In MacWhinney, B., editor, *Mechanisms of language acquisition*, pages 1–33. Erlbaum, Hillsdale, NJ.

Clark, E. V. and Cohen, S. R. (1984). Productivity and memory for newly formed words. *Journal of Child Language*, 11(3):611–625.

Clark, E. V. and Hecht, B. F. (1982). Learning to coin agent and instrument nouns. *Cognition*, 12(1):1–24.

Coady, J. A. and Aslin, R. N. (2004). Young children's sensitivity to probabilistic phonotactics in the developing lexicon. *Journal of Experimental Child Psychology*, 89(3):183–213.

Cohen, W. W. (1995). Fast effective rule induction. In *Proceedings of the Twelfth International Conference on Machine Learning, Lake Tahoe, California*.

Cohn, A. (2006). Is there gradient phonology? In Fanselow, G., Féry, C., Vogel, R., and Schlesewsky, M., editors, *Gradience in grammar: Generative perspectives*, pages 25–44. Oxford University Press, Oxford.

Colé, P., Beauvillain, C., and Segui, J. (1989). On the representation and processing of prefixed and suffixed derived words: A differential frequency effect. *Journal of Memory and language*, 28(1):1–13

Coleman, J. and Pierrehumbert, J. (1997). Stochastic phonological grammars and acceptability. In Coleman, J., editor, *Proceedings of the 3rd meeting of the ACL Special Interest Group in Computational Phonology*, pages 49–56, Somerset, NJ. Association for Computational Linguistics.

Collins, M. (1999). *Head-driven statistical models for natural language processing*. PhD thesis, University of Pennsylvania.

Conwell, E. and Demuth, K. (2007). Early syntactic productivity: Evidence from dative shift. *Cognition*, 103(2):163–179.

Cooreman, A. M. (1987). *Transitivity and discourse continuity in Chamorro narratives*. Walter de Gruyter, Berlin.

Coppock, E. (2008). *The logical and empirical foundations of Baker's Paradox*. PhD thesis, Stanford University.

Corbett, G. G. and Fraser, N. M. (1993). Network morphology: A DATR account of Russian nominal inflection. *Journal of Linguistics*, 29(1):113–142.

Cover, T. M. and Thomas, J. A. (2012). *Elements of information theory*. John Wiley & Sons, Hoboken, NJ.

Crain, S. (1991). Language acquisition in the absence of experience. *Behavioral and Brain Sciences*, 14(4):597–612.

Crain, S. and Thornton, R. (2000). *Investigations in universal grammar: A guide to experiments on the acquisition of syntax and semantics*. MIT Press, Cambridge, MA.

Culicover, P. W. (1999). *Syntactic nuts: Hard cases, syntactic theory, and language acquisition*. Oxford University Press, Oxford.

Cutler, A. (1996). Prosody and the word boundary problem. *Signal to syntax: Bootstrapping from speech to grammar in early acquisition*, pages 87–99.

Cutler, A. and Carter, D. M. (1987). The predominance of strong initial syllables in the English vocabulary. *Computer Speech and Language*, 2(3–4):133–142.

Dąbrowska, E. (2001). Learning a morphological system without a default: The Polish genitive. *Journal of Child Language*, 28(3):545–574.

Dąbrowska, E. (2005). Productivity and beyond: mastering the Polish genitive inflection. *Journal of Child Language*, 32(1):191–205.

Dąbrowska, E. and Szczerbinski, M. (2006). Polish children's productivity with case marking: the role of regularity, type frequency, and phonological diversity. *Journal of Child Language*, 33:559–597.

Daelemans, W., Gillis, S., and Durieux, G. (1994). The acquisition of stress: A data-oriented approach. *Computational Linguistics*, 20(3):421–451.

Daelemans, W., Zavrel, J., van der Sloot, K., and van den Bosch, A. (2009). TiMBL: Tilburg Memory-Based Learner. Technical Report ILK 09-01, Induction of Linguistic Knowledge Research Group, Tilburg University.

Dale, P. S. and Fenson, L. (1996). Lexical development norms for young children. *Behavior Research Methods*, 28(1):125–127.

Davies, M. (2008). Corpus of Contemporary American English (COCA): 410+ million words, 1990-present. Available at http://www.americancorpus.org/.

Davies, M. (2010). The Corpus of Historical American English: 400 million words, 1810-2009. Available at http://corpus.buy.edu/coha/.

Deen, K. U. (2005). *The acquisition of Swahili*, volume 40. John Benjamins Publishing, Amsterdam.

Demuth, K. (1996). The prosodic structure of early words. In Morgan, J. L. and Demuth, K., editors, *Signal to syntax: Bootstrapping from speech to grammar in early acquisition*, pages 171–186. Psychology Press.

Demuth, K. (2003). The acquisition of Bantu languages. In *The Bantu languages*, pages 209–222. Curzon Press Surrey,, United Kingdom.

Downing, P. A. (1996). *Numeral classifier systems: The case of Japanese*. John Benjamins Publishing, Amsterdam.

Dresher, B. E. and Kaye, J. (1990). A computational learning model for metrical phonology. *Cognition*, 34:137–195.

Dressler, W. U. (1999). Why collapse morphological concepts? *Behavioral and Brain Sciences*, 22(6):1021.

Dromi, E. (1987). *Early lexical development*. Cambridge University Press, Cambridge.

Echols, C. H., Crowhurst, M. J., and Childers, J. B. (1997). The perception of rhythmic units in speech by infants and adults. *Journal of memory and language*, 36(2):202–225.

Eddington, D. (1996). Diphthongization in Spanish derivational morphology: An empirical investigation. *Hispanic Linguistics*, 8(1):1–13.

Ellegård, A. (1953). *The auxiliary do: The establishment and regulations of its use in English*. Gothenberg Studies in English. Almqvist and Wiksell, Stockholm.

Elman, J. L. (1993). Learning and development in neural networks: The importance of starting small. *Cognition*, 48(1):71–99.

Elsen, H. (2002). The acquisition of German plurals. In *Morphology 2000: Selected Papers from the 9th Morphology Meeting, Vienna, 25-27 February 2000*, volume 218, page 117. John Benjamins Publishing.

Embick, D. and Halle, M. (2005). On the status of stems in morphological theory. In Geerts, T., van Ginneken, I., and Jacobs, H., editors, *Romance Languages and Linguistic Theory 2003: Selected papers from Going Romance 2003*, pages 37–62. John Benjamins, Amsterdam.

Embick, D. and Marantz, A. (2008). Architecture and blocking. *Linguistic Inquiry*, 39(1):1–53.

Eythórsson, T. (2002). Changes in subject case marking in Icelandic. In Lightfoot, D., editor, *Syntactic effects of morphological change*, pages 196–212. Oxford University Press, Oxford.

Fanselow, G. and Féry, C. (2002). Ineffability in grammar. In Fanselow, G. and Féry, C., editors, *Resolving conflicts in grammars: Optimality Theory in syntax, morphology, and phonology*, pages 265–307. Helmut Buske, Hamburg.

Feldman, J. (2000). Minimization of boolean complexity in human concept learning. *Nature*, 407(6804):630–633.

Feldman, N. H., Griffiths, T. L., Goldwater, S., and Morgan, J. L. (2013). A role for the developing lexicon in phonetic category acquisition. *Psychological Review*, 120(4):751–778.

Fenson, L., Dale, P. S., Reznick, J. S., Bates, E., Thal, D. J., Pethick, S. J., Tomasello, M., Mervis, C. B., and Stiles, J. (1994). Variability in early communicative development. *Monographs of the Society for Research in Child Development*, pages i–185.

Fikkert, P. (1994). *On the acquisition of prosodic structure*. PhD thesis, Leiden University.

Firth, J. (1957). A synopsis of linguistic theory 1930–1955. In Palmer, F. R., editor, *Selected papers of J.R. Firth*, pages 1–32. Longman, London.

Fleischhauer, E. and Clahsen, H. (2012). Generating inflected word forms in real time: Evaluating the role of age, frequency, and working memory. In Biller, A., Chung, E., and Kimball, A., editors, *Proceedings of the 36th annual Boston University Conference on Language Development*, volume 1, pages 164–176, Somerville, MA. Cascadilla Press.

Fodor, J. A., Bever, T. G., and Garrett, M. F. (1974). *The psychology of language*. McGraw-Hill, New York.

Fodor, J. D. (2001). Parameters and the periphery: Reflections on syntactic nuts. *Journal of Linguistics*, 37(2):367–392.

Fodor, J. D. and Crain, S. (1987). Simplicity and generality of rules in language acquisition. In MacWhinney, B., editor, *Mechanisms of language acquisition*, pages 35–63. Lawrence Erlbaum Associates, Hillsdale, NJ.

Fodor, J. D. and Sakas, W. G. (2005). The subset principle in syntax: Costs of compliance. *Journal of Linguistics*, 41(3):513–569.

Forster, K. I. (1976). Accessing the mental lexicon. In Wales, R. and Walker, E., editors, *New approaches to language mechanisms*, pages 257–287. North-Holland, Amsterdam.

Forster, K. I. (1992). Memory-addressing mechanisms and lexical access. In Forst, R. and Katz, L., editors, *Orthography, phonology, morphology, and meaning*, pages 413–434. Elsevier, Amsterdam.

Frank, M. C., Goodman, N. D., and Tenenbaum, J. B. (2009). Using speakers' referential intentions to model early cross-situational word learning. *Psychological Science*, 20(5):578–585.

Fraser, B. (1974). *The verb-particle combination in English*. Taishukan, Tokyo.

Fruchter, J. and Marantz, A. (2015). Decomposition, lookup, and recombination: MEG evidence for the Full Decomposition model of complex visual word recognition. *Brain and Language*, 143:81–96.

Fruchter, J., Stockall, L., and Marantz, A. (2013). MEG masked priming evidence for form-based decomposition of irregular verbs. *Frontiers in Human Neuroscience*, 7(798).

Gagliardi, A. and Lidz, J. (2014). Statistical insensitivity in the acquisition of Tsez noun classes. *Language*, 90(1):58–89.

Gawlitzek-Maiwald, I. (1994). How do children cope with variation in the input? the case of German plurals and compounding. In Tracy, R. and Lattey, E., editors, *How tolerant is Universal Grammar? Essays on language learnability and language variation*, pages 225–266. Niemeyer, Tübingen.

Gerken, L. (1994). Young children's representation of prosodic phonology: Evidence from English-speakers' weak syllable productions. *Journal of Memory and Language*, 33(1):19–38.

Gerken, L. and McIntosh, B. J. (1993). Interplay of function morphemes and prosody in early language. *Developmental Psychology*, 29(3):448–457.

Gerken, L., Wilson, R., and Lewis, W. (2005). Infants can use distributional cues to form syntactic categories. *Journal of Child Language*, 32(2):249–268.

Gibbs, R. W. (1980). Spilling the beans on understanding and memory for idioms in conversation. *Memory & Cognition*, 8(2):149–156.

Gibbs, R. W. and Gonzales, G. P. (1985). Syntactic frozenness in processing and remembering idioms. *Cognition*, 20(3):243–259.

Gibson, E. and Wexler, K. (1994). Triggers. *Linguistic Inquiry*, 25(3):407–454.

Gildea, D. (2001). Corpus variation and parser performance. In *Proceedings of the 2001 Conference on Empirical Methods in Natural Language Processing*, pages 167–202.

Gillette, J., Gleitman, H., Gleitman, L., and Lederer, A. (1999). Human simulations of vocabulary learning. *Cognition*, 73(2):135–176.

Gleitman, L. (1990). The structural sources of verb meanings. *Language Acquisition*, 1(1):3–55.

Gold, E. M. (1967). Language identification in the limit. *Information and Control*, 10:447–474.

Goldberg, A. E. (1995). *Constructions*. University of Chicago Press, Chicago.

Goldberg, A. E. (2003). Constructions: a new theoretical approach to language. *Trends in Cognitive Sciences*, 7(5):219–224.

Goldberg, A. E. (2011). Are a-adjectives like afraid prepositional phrases underlying and does it matter from a learnability perspective? Unpublished manuscript. Princeton University.

Goldsmith, J. (2001). Unsupervised learning of morphology of a natural language. *Computational Linguistics*, 27(2):153–198.

Goldwater, S., Griffiths, T. L., and Johnson, M. (2009). A bayesian framework for word segmentation: Exploring the effects of context. *Cognition*, 112(1):21–54.

Golinkoff, R. M., Hirsh-Pasek, K., Cauley, K. M., and Gordon, L. (1987). The eyes have it: Lexical and syntactic comprehension in a new paradigm. *Journal of Child Language*, 14(01):23–45.

Goodman, N. (1955). *Fact, fiction, and forecast*. Harvard University Press, Cambridge, MA.

Gordon, M. (2002). A factorial typology of quantity-insensitive stress. *Natural Language & Linguistic Theory*, 20(3):491–552.

Gorman, K. (2013). *Generative phonotactics*. PhD thesis, University of Pennsylvania.

Green, G. M. (1974). *Semantics and syntactic regularity*. Indiana University Press.

Grimshaw, J. (1990). *Argument structure*. MIT Press, Cambridge, MA.

Gropen, J., Pinker, S., Hollander, M., Goldberg, R., and Wilson, R. (1989). The learnability and acquisition of the dative alternation in English. *Language*, 65(2):203–257.

Gross, M. (1975). *Méthodes en syntaxe: Régime des constructions complétives*, volume 1365. Hermann, Paris.

Gross, M. (1979). On the failure of generative grammar. *Language*, 55(4):859–885.

Guasti, M. T. (1993). Verb syntax in Italian child grammar: Finite and nonfinite verbs. *Language Acquisition*, 3(1):1–40.

Guion, S., Clark, J., Harada, T., and Wayland, R. (2003). Factors affecting stress placement for English nonwords include syllabic structure, lexical class, and stress patterns of phonologically similar words. *Language and Speech*, 46(4):403–427.

Gxilishe, S., de Villiers, P., de Villiers, J., Belikova, A., Meroni, L., and Umeda, M. (2007). The acquisition of subject agreement in Xhosa. In *Proceedings of the Conference on Generative Approaches to Language Acquisition (GALANA)*, volume 2, pages 114–123.

Haegeman, L. (1990). Understood subjects in English diaries: On the relevance of theoretical syntax for the study of register variation. *Multilingua*, 9(2):157–199.

Hahn, U. and Nakisa, R. C. (2000). German inflection: Single route or dual route? *Cognitive Psychology*, 41(4):313–360.

Hale, K. and Keyser, S. J. (2002). *Prolegomenon to a theory of argument structure*. MIT Press, Cambridge, MA.

Halldórsson, H. (1982). On dative substitution. *Íslenskt mál og almenn málfræði*, 4:159–189.

Halle, M. (1962). Phonology in generative grammar. *Word*, 18(1):54–72.

Halle, M. (1973). Prolegomena to a theory of word formation. *Linguistic Inquiry*, 4(1):3–16.

Halle, M. (1978). Knowledge unlearned and untaught: What speakers know about the sounds of their language. In Halle, M., Bresnan, J., and Miller, G. A., editors, *Linguistic theory and psychological reality*, pages 294–303. MIT Press, Cambridge, MA.

Halle, M. (1997). Distributed morphology: Impoverishment and fission. *MIT Working Papers in Linguistics*, 30:425–449.

Halle, M. (1998). The stress of English words 1968-1998. *Linguistic Inquiry*, 29(4):539–568.

Halle, M. and Marantz, A. (1993). Distributed morphology and the pieces of inflection. In Hale, K. and Keyser, S. J., editors, *The view from Building 20: Essays in linguistics in honor of Sylvain Bromberger*, pages 111–176. MIT Press, Cambridge, MA.

Halle, M. and Marantz, A. (1994). Some key features of Distributed Morphology. *MIT Working Papers in Linguistics*, 21:275–288.

Halle, M. and Mohanan, K. P. (1985). Segmental phonology of Modern English. *Linguistic Inquiry*, 16(1):57–116.

Halle, M. and Vergnaud, J.-R. (1987). *An essay on stress*. MIT Press, Cambridge, MA.

Hankamer, J. (1992). Morphological parsing and the lexicon. In Marslen-Wilson, W. D., editor, *Lexical representation and process*, pages 392–408. MIT Press, Cambridge, MA.

Harley, H. (2002). Possession and the double object construction. *Language Variation Yearbook*, 2:29–68.

Harley, H. (2014). On the identity of roots. *Theoretical linguistics*, 40(3–4):225–276.

Harris, J. W. (1969). *Spanish phonology*. MIT Press, Cambridge, MA.

Harris, J. W. (1977). Remarks of diphthongization in Spanish. *Lingua*, 41(3–4):261–305.

Harris, Z. S. (1951). *Methods in structural linguistics*. University of Chicago Press, Chicago.

Harris, Z. S. (1954). Distributional structure. *Word*, 10(23):146–162.

Harris, Z. S. (1955). From phoneme to morpheme. *Language*, 31(2):190–222.

Hart, B. and Risley, T. R. (1995). *Meaningful differences in the everyday experience of young American children*. Paul H Brookes Publishing, Baltimore, MD.

Hart, B. and Risley, T. R. (2003). The early catastrophe: The 30 million word gap by age 3. *American Educator*, 27(1):4–9.

Hauser, M. D., Chomsky, N., and Fitch, W. T. (2002). The faculty of language: What is it, who has it, and how did it evolve? *Science*, 298(5598):1569–1579.

Hauser, M. D., Yang, C., Berwick, R. C., Tattersall, I., Ryan, M. J., Watumull, J., Chomsky, N., and Lewontin, R. C. (2014). The mystery of language evolution. *Frontiers in Psychology*, 5:doi: 10.3389/fpsyg.2014.00401.

Hay, J. and Baayen, R. H. (2005). Shifting paradigms: Gradient structure in morphology. *Trends in Cognitive Sciences*, 9(7):342–348.

Hayes, B. (1982). Extrametricality and English stress. *Linguistic Inquiry*, 13(2):227–276.

Hayes, B. (1995). *Metrical stress theory*. University of Chicago Press, Chicago.

Hayes, B. and Wilson, C. (2008). A maximum entropy model of phonotactics and phonotactic learning. *Linguistic Inquiry*, 39(3):379–440.

Hayes, B., Zuraw, K., Siptár, P., and Londe, Z. (2009). Natural and unnatural constraints in Hungarian vowel harmony. *Language*, 85(4):822–863.

Heath, S. B. (1983). *Ways with words: Language, life and work in communities and classrooms*. Cambridge University Press, Cambridge.

Herman, L. and Herman, M. S. (2014). *American dialects: A manual for actors, directors, and writers*. Routledge, New York.

Hetzron, R. (1975). Where the grammar fails. *Language*, 51(4):859–872.

Hockett, C. F. (1942). English verb inflection. *Studies in Linguistics*, 1(2):1–8.

Holmberg, A. (2010). Parameters in minimalist theory: The case of Scandinavian. *Theoretical Linguistics*, 36(1):1–48.

Horn, L. (1989). *A natural history of negation*. University of Chicago Press, Chicago.

Hornstein, N. (2009). *A theory of syntax: Minimal operations and universal grammar*. Cambridge University Press, Cambridge.

Hovav, M. R. and Levin, B. (2008). The English dative alternation: The case for verb sensitivity. *Journal of Linguistics*, 44(1):129.

Howes, D. H. and Solomon, R. L. (1951). Visual duration threshold as a function of word-probability. *Journal of Experimental Psychology*, 41(6):401–410.

Hu, Q. (1993). Overextension of animacy in Chinese classifier acquisition. In *The proceedings of the twenty-fifth annual child language research forum*, volume 25, page 127, Stanford, CA. Center for the Study of Language (CSLI).

Huang, C.-T. J. (1984). On the distribution and reference of empty pronouns. *Linguistic inquiry*, 15(4):531–574.

Huddleston, R. and Pullum, G. K. (2001). *The Cambridge grammar of the English language*. Cambridge University Press, Cambridge.

Hudson, R. (1997). The rise of auxiliary do-verb raising or category-strengthening? *Transactions of the Philological Society*, 95(1):41–72.

Hudson, R. (2000). *I amn't. *Language*, 76(2):297–323.

Hudson Kam, C. L. and Newport, E. L. (2005). Regularizing unpredictable variation: The roles of adult and child learners in language formation and change. *Language Learning and Development*, 1(2):151–195.

Hudson Kam, C. L. and Newport, E. L. (2009). Getting it right by getting it wrong: When learners change languages. *Cognitive psychology*, 59(1):30–66.

Hurford, J. R. (2011). *The origins of grammar: Language in the light of evolution*, volume 2. Oxford University Press, Oxford.

Huttenlocher, J., Haight, W., Bryk, A., Seltzer, M., and Lyons, T. (1991). Early vocabulary growth: Relation to language input and gender. *Developmental Psychology*, 27(2):236.

Huttenlocher, J. and Smiley, P. (1987). Early word meanings: The case of object names. *Cognitive Psychology*, 19(1):63–89.

Hyams, N. (1986). *Language acquisition and the theory of parameters*. Reidel, Dordrecht.

Idsardi, W. J. (1992). *The computation of prosody*. PhD thesis, MIT.

Inkelas, S. and Zoll, C. (2005). *Reduplication: Doubling in morphology*. Cambridge University Press, Cambridge.

Itô, J. and Mester, A. (2003). *Japanese morphophonemics: Markedness and word structure*. MIT Press, Cambridge, MA.

Jackendoff, R. S. (1975). Morphological and semantic regularities in the lexicon. *Language*, 51(3):639–671.

Jackendoff, R. S. (1990a). On Larson's treatment of the double object construction. *Linguistic Inquiry*, 21(3):427–456.

Jackendoff, R. S. (1990b). *Semantic structures*. MIT Press, Cambridge, MA.

Jackendoff, R. S. (1996). *The architecture of the language faculty*. MIT Press, Cambridge, MA.

Jackendoff, R. S. (2007). *Language, consciousness, culture: Essays on mental structure*. MIT Press, Cambridge, MA.

Jaeggli, O. and Safir, K. J. (1989). The null subject parameter and parametric theory. In *The null subject parameter*, pages 1–44. Springer, Berlin.

Janda, R. D. (1990). Frequency, markedness and morphological change: On predicting the spread of noun-plural-s in Modern High German and West Germanic. In *ESCOL*, volume 90, pages 136–153.

Jarmulowicz, L. (2002). English derivational suffix frequency and children's stress judgements. *Brain and Language*, 81(1–3):192–204.

Jelinek, E. and Carnie, A. (2003). Argument hierarchies and the mapping principle. In Carnie, A., Harley, H., and Willie, M., editors, *Formal approaches to function in grammar*, pages 265–296. John Benjamins, Amsterdam.

Jelinek, F. (1998). *Statistical methods for speech recognition*. MIT Press, Cambridge, MA.

Jespersen, O. (1942). *A Modern English grammar on historical principles. Part VI: Morphology*. Ejnar Munksgaard, Copenhagen.

Johnson, E. K. and Jusczyk, P. W. (2001). Word segmentation by 8-month-olds: When speech cues count more than statistics. *Journal of Memory and Language*, 44(4):493–666.

Johnson, M., Griffiths, T. L., and Goldwater, S. (2006). Adaptor grammars: A framework for specifying compositional nonparametric bayesian models. In *Advances in Neural Information Processing Systems*, pages 641–648.

Jónsson, J. G. (1997). Verbs taking oblique subjects. *Íslenskt mál og almenn málfræði*, 19-20:11–43.

Jónsson, J. G. (2003). Not so quirky: On subject case in Icelandic. In Brandner, E. and Zinsmeister, H., editors, *New Perspectives in Case Theory*, pages 127–163. Center for the Study of Language and Information, Stanford, CA.

Jónsson, J. G. and Eythórsson, T. (2005). Variation in subject case marking in Insular Scandinavian. *Nordic Journal of Linguistics*, 28(02):223–245.

Jónsson, J. G. and Eythórsson, T. (2011). Structured exceptions and case selection in Insular Scandinavian. In Simon, H. J. and Wiese, H., editors, *Expecting the unexpected: Exceptions in grammar*, volume 216, pages 213–241. Walter de Gruyter, Berlin.

Jung, Y.-J. and Miyagawa, S. (2004). Decomposing ditransitive verbs. In *Proceedings of SICGG*, pages 101–120.

Jusczyk, P. W. and Aslin, R. N. (1995). Infant's detection of the sound patterns of words in fluent speech. *Cognitive Psychology*, 46(1):65–97.

Jusczyk, P. W., Cutler, A., and Redanz, N. J. (1993a). Infants' preference for the predominant stress patterns of English words. *Child Development*, 64(3):675–687.

Jusczyk, P. W., Friederici, A. D., Wessels, J. E., Svenkerud, V., and Jusczyk, A. (1993b). Infants' sensitivity to the sound patterns of native language words. *Journal of Memory and Language*, 32(3):402–420.

Jusczyk, P. W. and Hohne, E. A. (1997). Infants' memory for spoken words. *Science*, 277:1984–1986.

Jusczyk, P. W., Houston, D. M., and Newsome, M. (1999). The beginnings of word segmentation in English-learning infants. *Cognitive Psychology*, 39(3-4):159–207.

Kahn, D. (1976). *Syllable-based generalizations in English phonology*. PhD thesis, MIT. Published by Garland, New York, 1980.

Kehoe, M. (1997). Stress error patterns in English-speaking children's word productions. *Clinical Linguistics & Phonetics*, 11(5):389–409.

Kehoe, M. and Stoel-Gammon, C. (1997). The acquisition of prosodic structure: An investigation of current accounts of children's prosodic development. *Language*, 73:113–144.

Kelly, M. H. (1992). Using sound to solve syntactic problems: The role of phonology in grammatical category assignments. *Psychological Review*, 99(2):349–364.

Kemp, C., Perfors, A., and Tenenbaum, J. B. (2007). Learning overhypotheses with hierarchical bayesian models. *Developmental Science*, 10(3):307–321.

Keshava, S. and Pitler, E. (2006). A simpler, intuitive approach to morpheme induction. In *Proceedings of 2nd Pascal Challenges Workshop*, pages 31–35.

Kiparsky, P. (1965). *Phonological change*. PhD thesis, MIT.

Kiparsky, P. (1966). Über den deutschen Akzent. *Studia Grammatica*, 7:69–97.

Kiparsky, P. (1973). Elsewhere in phonology. In Anderson, S. R. and Kiparsky, P., editors, *A festschrift for Morris Halle*, pages 93–106. Holt, Rinehart and Winston, New York.

Kiparsky, P. (1974). Remarks on analogical change. In Anderson, J. M. and Jones, C., editors, *Historical linguistics II: Theory and description in phonology*, pages 257–276. North-Holland, Amsterdam.

Kiparsky, P. (1982). Lexical morphology and phonology. In Yang, I.-S., editor, *Linguistics in the Morning Calm: Selected Papers from SICOL 1981*, pages 3–91. Hanshin, Seoul.

Köpcke, K.-M. (1998). The acquisition of plural marking in English and German revisited: Schemata versus rules. *Journal of Child Language*, 25(2):293–319.

Kottum, S. E. (1981). The genitive singular form of masculine nouns in Polish. *Scando-Slavica*, 27(1):179–186.

Krifka, M. (1999). Manner in dative alternation. In *West Coast Conference on Formal Linguistics*, volume 18, pages 260–271.

Kroch, A. (1989). Reflexes of grammar in patterns of language change. *Language Variation and Change*, 1(3):199–244.

Kučera, H. and Francis, W. N. (1967). *Computational analysis of present-day American English*. Brown University Press, Providence.

Kuczaj, S. A. (1976). *-ing, -s, and -ed: A study of the acquisition of certain verb inflections*. PhD thesis, University of Minnesota.

Kuczaj, S. A. (1977). The acquisition of regular and irregular past tense forms. *Journal of Verbal Learning and Verbal Behavior*, 16(5):589–600.

Kuhl, P. K., Williams, K. A., Lacerda, F., Stevens, K. N., and Lindblom, B. (1992). Linguistic experience alters phonetic perception in infants by 6 months of age. *Science*, 255(5044):606–608.

Laaha, S., Ravid, D., Korecky-Kröll, K., Laaha, G., and Dressler, W. U. (2006). Early noun plurals in German: Regularity, productivity or default? *Journal of Child Language*, 33(2):271–302.

Labov, W. (1972a). *Language in the inner city: Studies in Black English Vernacular.* University of Pennsylvania Press, Philadelphia.

Labov, W. (1972b). *Sociolinguistic patterns.* University of Pennsylvania Press, Philadelphia.

Labov, W. (1989). The child as linguistic historian. *Language Variation and Change,* 1(1):85–97.

Labov, W. (1994). *Principles of linguistic change: Internal factors.* Blackwell, Oxford.

Labov, W. (2007). Transmission and diffusion. *Language,* 83(2):344–387.

Labov, W., Ash, S., and Boberg, C. (2006). *The atlas of North American English: Phonetics, phonology, and sound change.* Mouton de Gruyter, Berlin.

Ladefoged, P. and Fromkin, V. (1968). Experiments on competence and performance. *IEEE Transactions on Audio and Electroacoustics,* 16(1):130–136.

Lakoff, G. (1970). *Irregularity in syntax.* Holt, Rinehart and Winston, New York.

Landau, B. and Gleitman, L. R. (1984). *Language and experience: Evidence from the blind child,* volume 8. Harvard University Press, Cambridge, MA.

Larson, R. K. (1988). On the double object construction. *Linguistic Inquiry,* 19(3):335–391.

Larson, R. K. (1990). Double objects revisited: Reply to Jackendoff. *Linguistic Inquiry,* 21(4):589–632.

Larson, R. K. and Marušič, F. (2004). On indefinite pronoun structures with aps: Reply to Kishimoto. *Linguistic Inquiry,* 35(2):268–287.

Legate, J. A. (2008). Morphological and abstract case. *Linguistic Inquiry,* 39(1):55–101.

Legate, J. A. and Yang, C. (2002). Empirical re-assessment of stimulus poverty arguments. *The Linguistic Review,* 19:151–162.

Legate, J. A. and Yang, C. (2007). Morphosyntactic learning and the development of tense. *Language Acquisition,* 14(3):315–344.

Legate, J. A. and Yang, C. (2013). Assessing child and adult grammar. In Berwick, R. and Piattelli-Palmarini, M., editors, *Rich languages from poor inputs: In honor of Carol Chomsky,* pages 168–182. Oxford University Press, Oxford.

Levelt, W. J., Roelofs, A., and Meyer, A. S. (1999). A theory of lexical access in speech production. *Behavioral and Brain Sciences,* 22(1):1–75.

Levin, B. (1993). *English Verb Classes and Alternations: A Preliminary Investigation.* University of Chicago Press, Chicago.

Levin, B. (2008). Dative verbs: A crosslinguistic perspective. *Lingvisticæ Investigationes,* 31(2):285–312.

Levine, M. (1975). *A cognitive theory of learning: Research on hypothesis testing.* Lawrence Erlbaum, Hillsdale, NJ.

Levy, Y. and Vainikka, A. (2000). The development of a mixed null subject system: a crosslinguistic perspective with data on the acquisition of Hebrew. *Language Acquisition,* 8(4):363–384.

Li, W. (1992). Random texts exhibit Zipf-law-like word frequency distributions. *IEEE Transactions in Information Theory,* 38(6):1842–1845.

Liberman, M. and Prince, A. (1977). On stress and linguistic rhythm. *Linguistic Inquiry,* 8:249–336.

Lieber, R. (1980). *On the organization of the lexicon.* PhD thesis, MIT.

Lieber, R. (1981). Morphological conversion within a restrictive theory of the lexicon. In Moortgat, M., van der Hulst, H., and Hoekstra, T., editors, *The scope of lexical rules,* pages 161–200. Foris.

Lightfoot, D. W. (1979). *Principles of diachronic syntax.* Cambridge University Press, Cambridge.

Lignos, C. (2010). Learning from unseen data. In *Proceedings of the Morpho Challenge 2010 Workshop*, pages 35–38.

Lignos, C. (2011). Modeling infant word segmentation. In *Proceedings of the 15th Conference on Computational Language Learning*, pages 28–38.

Lignos, C. (2013). *Modeling words in the mind*. PhD thesis, University of Pennsylvania.

Lignos, C., Chan, E., Marcus, M., and Yang, C. (2009). A rule-based unsupervised morphology learning framework. In *Working Notes for the CLEF 2009 Workshop*.

Lignos, C., Chan, E., Marcus, M., and Yang, C. (2010). Evidence for a morphological acquisition model from development data. In Franich, K., Iserman, K. M., and Keil, L. L., editors, *Proceedings of the 34th annual Boston University conference on language development*, volume 2, pages 269–280.

Lignos, C. and Gorman, K. (2012). Revisiting frequency and storage in morphological processing. *Proceedings of CLS*, 48:447–461.

Long, R. B. (1969). *The sentence and its parts*. University of Chicago Press, Chicago.

MacWhinney, B. (1975). Rules, rote, and analogy in morphological formations by Hungarian children. *Journal of Child Language*, 19(2):65–77.

MacWhinney, B. (2000). *The CHILDES project: Tools for analyzing talk*. Lawrence Erlbaum, Mahwah, NJ, 3rd edition.

Maiden, M. and O'Neill, P. (2010). On morphomic defectiveness: Evidence from the Romance languages of the Iberian Peninsula. In Baerman, M., Corbett, G. G., and Brown, D., editors, *Defective paradigms: Missing forms and what they tell us*, pages 103–124. Oxford University Press, Oxford.

Malkiel, Y. (1966). Diphthongization, monothongization, metaphony: studies in their interaction in the paradigm of the Old Spanish -IR verbs. *Language*, 42(2):430–472.

Mandelbrot, B. (1954). Structure formelle des textes et communication. *Word*, 10(1):1–27.

Marantz, A. (1997). No escape from syntax: Don't try morphological analysis in the privacy of your own lexicon. In Dimitriadis, A., Siegel, L., Surek-Clark, C., and Williams, A., editors, *Penn Working Papers in Linguistics 4.2: Proceedings of the 21st annual Penn Linguistics Colloquium*, pages 201–225. Penn Linguistics Club, Philadelphia.

Marantz, A. (2000). Case and licensing. In Reuland, E. J., editor, *Arguments and case: Explaining Burzio's generalization*, pages 11–30. John Benjamins, Amsterdam.

Marantz, A. (2001). Words. In *20th West Coast Conference on Formal Linguistics, University of Southern California*.

Marantz, A. (2013). Locality domains for contextual allomorphy across the interfaces. In Matushansky, O. and Marantz, A., editors, *Distributed Morphology today: Morphemes for Morris Halle*, pages 95–115. MIT Press, Cambridge, MA.

Maratsos, M. (2000). More overregularizations after all: New data and discussion on Marcus, Pinker, Ullman, Hollander, Rosen and Xu. *Journal of Child Language*, 27:183–212.

Maratsos, M. and Chalkley, M. A. (1980). The internal language of children's syntax: The nature and ontogenesis of syntactic categories. In Nelson, K., editor, *Children's language*, volume 2, pages 127–214. Gardner, Cincinnati, OH.

Marchand, H. (1969). *The categories and types of present-day English word-formation: A sychronic-diachronic approach*. C.H Beck'sche, München, 2nd edition.

Marcus, G., Pinker, S., Ullman, M. T., Hollander, M., Rosen, J., and Xu, F. (1992). *Overregularization in language acquisition*. Monographs of the Society for Research in Child Development. University of Chicago Press, Chicago.

Marcus, G. F. (1993). Negative evidence in language acquisition. *Cognition*, 46(1):53–85.

Marcus, G. F. (1995). The acquisition of the English past tense in children and multilayered connectionist networks. *Cognition*, 56(3):271–279.

Marcus, G. F., Brinkmann, U., Clahsen, H., Wiese, R., and Pinker, S. (1995). German inflection: The exception that proves the rule. *Cognitive Psychology*, 29:189–256.

Marcus, M. P., Santorini, B., Marcinkiewicz, M. A., and Taylor, A. (1999). Treebank-3. Linguistic Data Consortium: LDC99T42.

Marslen-Wilson, W. D. (1987). Functional parallelism in spoken word-recognition. *Cognition*, 25(1–2):71–102.

Marslen-Wilson, W. D. and Tyler, L. K. (1997). Dissociating types of mental computation. *Nature*, 387:592–594.

Maslen, R., Theakston, A. L., Lieven, E. V., and Tomasello, M. (2004). A dense corpus study of past tense and plural overregularization in English. *Journal of Speech, Language and Hearing Research*, 47(6):1319–1333.

Mateo, F. and Rojo Sastre, A. (1995). *El arte de conjugar en español*. Hatier, Paris.

Matthews, P. H. (1974). *Morphology: An introduction to the theory of word-structure*. Cambridge University Press, Cambridge.

Mausch, H. (2003). Current alternations in inflection of Polish masculine inanimate nouns in the singular: A pilot study. *Investigationes Linguisticae*, 9:1–21.

Mayol, L. (2007). Acquisition of irregular patterns in Spanish verbal morphology. In Nurmi, V. and Sustretov, D., editors, *Proceedings of the twelfth ESSLLI Student Session*, pages 1–11, Dublin.

Mazurkewich, I. (1984). The acquisition of the dative alternation by second language learners and linguistic theory. *Language Learning*, 34(1):91–108.

Mazurkewich, I. and White, L. (1984). The acquisition of the dative alternation: Unlearning overgeneralizations. *Cognition*, 16(3):261–283.

McCarthy, J. J. and Prince, A. (1995). Faithfulnesss and reduplicative identity. In Beckman, J., Dickey, L. W., and Urbancyzk, S., editors, *Papers in Optimality Theory*, pages 249–384. GLSA, Amherst, MA.

McClelland, J. L. and Bybee, J. (2007). Gradience of gradience: A reply to Jackendoff. *The Linguistic Review*, 24(4):437–455.

McClelland, J. L. and Patterson, K. (2002). Rules or connections in past-tense inflections: What does the evidence rule out? *Trends in Cognitive Sciences*, 6(11):465–472.

McClelland, J. L. and Rumelhart, D. (1981). An interactive activation model of context effects on letter perception: Part 1. an account of the basic findings. *Psychological Review*, 88(?):375–407.

McClosky, D., Charniak, E., and Johnson, M. (2010). Automatic domain adaptation for parsing. In *Human Language Technologies: The 2010 Annual Conference of the North American Chapter of the Association for Computational Linguistics*, pages 28–36. Association for Computational Linguistics.

McNeill, D. (1966). The creation of language by children. In Lyons, J. and Wales, R., editors, *Psycholinguistic Papers*, pages 99–132. Edinburgh University Press, Edinburgh.

Medin, D. L. and Schaffer, M. M. (1978). Context theory of classification learning. *Psychological Review*, 85(3):207.

Medina, T. N., Snedeker, J., Trueswell, J. C., and Gleitman, L. R. (2011). How words can and cannot be learned by observation. *Proceedings of the National Academy of Sciences*, 108(22):9014–9019.

Melançon, A. and Shi, R. (2015). Representations of abstract grammatical feature agreement in young children. *Journal of Child Language*, pages 1–15.

Mervis, C. B. and Johnson, K. E. (1991). Acquisition of the plural morpheme: A case study. *Developmental Psychology*, 27(2):222.

Mikolov, T., Chen, K., Corrado, G., and Dean, J. (2013). Efficient estimation of word representations in vector space. *arXiv preprint arXiv:1301.3781*.

Miller, G. A. (1957). Some effects of intermittent silence. *American Journal of Psychology*, 70(2):311–314.

Miller, G. A. (1991). *The science of words*. Scientific American Library, San Francisco.

Miller, G. A. and Chomsky, N. (1963). Finitary models of language users. In Luce, R. D., Bush, R. R., and Galanter, E., editors, *Handbook of mathematical psychology. Volume II*, pages 419–491. Wiley, New York.

Mills, A. (1986). *The acquisition of gender: A study of English and German*. Springer, Berlin.

Mitchell, T. M. (1982). Generalization as search. *Artificial Intelligence*, 18(2):203–226.

Molnar, R. A. (2001). "Generalize and sift" as a model of inflection acquisition. Master's thesis, MIT.

Morris, J. and Stockall, L. (2012). Early, equivalent ERP masked priming effects for regular and irregular morphology. *Brain and Language*, 123(2):81–93.

Mugdan, J. (1977). *Flexionsmorphologie und Psycholinguistik*. Narr, Tübingen.

Mulford, R. (1985). Comprehension of Icelandic pronoun gender: semantic versus formal factors. *Journal of Child Language*, 12(02):443–453.

Murray, W. S. and Forster, K. I. (2004). Serial mechanisms in lexical access: The rank hypothesis. *Psychological Review*, 111(3):721–756.

Myers, S. (1987). Vowel shortening in English. *Natural Language and Linguistic Theory*, 5(4):485–518.

Nagy, W. E. and Anderson, R. C. (1984). How many words are there in printed school English? *Reading Research Quarterly*, 19(3):304–330.

Naigles, L. (1990). Children use syntax to learn verb meanings. *Journal of Child Language*, 17(02):357–374.

Naigles, L. R. and Hoff-Ginsberg, E. (1998). Why are some verbs learned before other verbs? Effects of input frequency and structure on children's early verb use. *Journal of Child Language*, 25(01):95–120.

Neue, F. (1866). *Formenlehre der lateinischen Sprache*. H. Lindemann, Stuttgart.

New, B., Brysbaert, M., Segui, J., Ferrand, L., and Rastle, K. (2004). The processing of singular and plural nouns in French and English. *Journal of Memory and Language*, 51(4):568–585.

Newmeyer, F. J. (2004). Against a parameter-setting approach to typological variation. *Linguistic Variation Yearbook*, 4(1):181–234.

Newport, E. (1990). Maturational constraints on language learning. *Cognitive Science*, 14(1):11–28.

Niemi, J., Laine, M., and Tuominen, J. (1994). Cognitive morphology in finnish: Foundations of a new model. *Language and Cognitive Processes*, 9(3):423–446.

Niyogi, P. (2006). *The computational nature of language learning and evolution*. MIT Press, Cambridge, MA.

Niyogi, P. and Berwick, R. (1996). Formaling triggers: A learning model for finite space. *Linguistic Inquiry*, 27(4):605–622. Revised version of MIT AI Laboratory Memo #1449, 1993.

Nowenstein, I. E. (2015). Acquiring intra-speaker variation: The case of icelandic dative substitution. To appear in the Proceedings of the Generative Approach to Language Acquisition in North America (GALANA). University of Maryland, College Park.

Noyer, R. (1992). *Features, positions, and affixes in automous morphological structure*. PhD thesis, MIT. Published by Garland, New York, 1997.

O'Donnell, T. (2015). *Productivity and reuse in language*. MIT Press, Cambridge, MA.

Oehrle, R. T. (1976). *The grammatical status of the English dative alternation.* PhD thesis, Massachusetts Institute of Technology.

Oh, G., Guion-Anderson, S., and Redford, M. A. (2011). Developmental change in factors affecting stress placement in native English-speaking children. In *Proceedings of the 17th International Congress of Phonetic Sciences*, pages 1522–1525.

Orgun, C. O. and Sprouse, R. (1999). From MParse to CONTROL: Deriving ungrammaticality. *Phonology*, 16(2):191–224.

Osherson, D. N. and Smith, E. E. (1981). On the adequacy of prototype theory as a theory of concepts. *Cognition*, 9(1):35–58.

Osherson, D. N., Stob, M., and Weinstein, S. (1986). *Systems that learn: An introduction to learning theory for cognitive and computer scientists.* MIT Press, Cambridge, MA.

Parducci, A. and Perrett, L. F. (1971). Category rating scales: Effects of relative frequency of stimulus values. *Journal of Experimental Psychology*, 89(2):427–452.

Park, T. (1978). Plurals in child speech. *Journal of Child Language*, 5:237–250.

Paul, H. (1888). *Principles of the History of Language.* Swan Sonnenschein, London.

Pearl, L. and Sprouse, J. (2013). Syntactic islands and learning biases: Combining experimental syntax and computational modeling to investigate the language acquisition problem. *Language Acquisition*, 20(1):23–68.

Penke, M. and Krause, M. (2002). German noun plurals: A challenge to the dual-mechanism approach. *Brain and Language*, 81(1–3):301–311.

Penny, R. (2002). *A history of the Spanish language.* Cambridge University Press, Cambridge, 2nd edition.

Pérez-Pereira, M. (1991). The acquisition of gender: What Spanish children tell us. *Journal of Child Language*, 18(03):571–590.

Perfors, A., Tenenbaum, J. B., and Regier, T. (2011). The learnability of abstract syntactic principles. *Cognition*, 118(3):306–338.

Perfors, A., Tenenbaum, J. B., and Wonnacott, E. (2010). Variability, negative evidence, and the acquisition of verb argument constructions. *Journal of Child Language*, 37(3):607–642.

Pertsova, K. (2005). How lexical conservatism can lead to paradigm gaps. In Heinz, J., Martin, A., and Pertsova, K., editors, *UCLA Working Papers in Linguistics 11: Papers in Phonology 6*, pages 13–30. UCLA Linguistics Department, Los Angeles.

Pesetsky, D. (1977). Russian morphology and lexical theory. Ms., MIT.

Pesetsky, D. (1995). *Zero syntax: Experiencer and Cascade.* MIT Press, Cambridge, MA.

Peters, A. M. (1983). *The units of language acquisition.* Cambridge University Press, Cambridge.

Pica, T. (1983). Adult acquisition of English as a second language under different conditions of exposure. *Language learning*, 33(4):465–497.

Pierrehumbert, J. (2003). Probabilistic phonology: Discrimination and robustness. In Bod, R., Hay, J., and Jannedy, S., editors, *Probabilistic Linguistics*, pages 177–228. MIT Press, Cambridge, MA.

Pine, J. M. and Lieven, E. V. (1997). Slot and frame patterns and the development of the determiner category. *Applied Psycholinguistics*, 18(2):123–138.

Pinker, S. (1989). *Learnability and cognition: The acquisition of argument structure.* MIT Press, Cambridge, MA.

Pinker, S. (1991). Rules of language. *Science*, 253:530–535.

Pinker, S. (1995). Why the child holded the baby rabbit: A case study in language acquisition. In Gleitman, L. R. and Liberman, M., editors, *An invitation to cognitive science, Vol. 1: Language*, pages 107–133. MIT Press, Cambridge, MA.

Pinker, S. (1999). *Words and rules: The ingredients of language*. Basic Books, New York.

Pinker, S. and Prince, A. (1988). On language and connectionism: Analysis of a parallel distributed processing model of language acquisition. *Cognition*, 28(1):73–193.

Pinker, S. and Ullman, M. T. (2002). The past and future of the past tense. *Trends in Cognitive Science*, 6(11):456–463.

Pizzuto, E. and Caselli, M. C. (1994). The acquisition of Italian verb morphology in a crosslinguistic perspective. In Levy, Y., editor, *Other children, other languages*, pages 137–187. Lawrence Erlbaum, Hillsdale, NJ.

Plag, I. (2003). *Word-formation in English*. Cambridge University Press, Cambridge.

Plag, I. and Baayen, H. (2009). Suffix ordering and morphological processing. *Language*, 85(1):109–152.

Plag, I., Dalton-Puffer, C., and Baayen, R. H. (1999). Morphological productivity across speech and writing. *English Language and Linguistics*, 3(2):209–228.

Plaut, D. C. (1997). Structure and function in the lexical system: Insights from distributed models of word reading and lexical decision. *Language and Cognitive Processes*, 12(5):765–806.

Plunkett, K. and Marchman, V. (1991). U-shaped learning and frequency effects in a multi-layered perception: Implications for child language acquisition. *Cognition*, 38(1):43–102.

Prasada, S. and Pinker, S. (1993). Generalisation of regular and irregular morphological patterns. *Language and Cognitive Processes*, 8(1):1–56.

Prince, A. and Smolensky, P. (2004). *Optimality Theory: Constraint interaction in generative grammar*. MIT Press, Cambridge, MA.

Pullum, G. K. and Wilson, D. (1977). Autonomous syntax and the analysis of auxiliaries. *Language*, 53(4):741–788.

Quine, W. V. O. (1960). *Word and object*. MIT Press, Cambridge, MA.

Raffelsiefen, R. (1996). Gaps in word formation. In Kleinhenz, U., editor, *Interfaces in phonology*, pages 194–209. Akademie-Verlag, Berlin.

Raffelsiefen, R. (1999). Phonological constraints on English word fomation. In Booij, G. and van Marle, J., editors, *Yearbook of Morphology 1998*, pages 225–287. Kluwer, Dordrecht.

Raffelsiefen, R. (2004). Absolute ill-formedness and other morphophonological effects. *Phonology*, 21(1):91–142.

Ramscar, M. (2002). The role of meaning in inflection: Why the past tense does not require a rule. *Cognitive Psychology*, 45(1):45–94.

Randall, J. (1990). Catapults and pendulums: the mechanics of language acquisition. *Linguistics*, 28(6):1381–1406.

Randall, J. H. (1980). -ity: A study in word formation restrictions. *Journal of Psycholinguistic Research*, 9(6):523–534.

Randall, J. H. (1987). *Indirect positive evidence: Overturning overgeneralizations in language acquisition*. Indiana University Linguistics Club.

Rastle, K., Davis, M. H., and New, B. (2004). The broth in my brother's brothel: Morpho-orthographic segmentation in visual word recognition. *Psychonomic Bulletin and Review*, 11(6):1090–1098.

Rauh, G. (1993). On the grammar of lexical and non-lexical prepositions in English. In Aurnague, M. and Vieu, L., editors, *The semantics of prepositions*, pages 99–150. Walter de Gruyter, Berlin.

Real Academia Española (1992). *Diccionario de la lengua española*. Real Academia Española, Madrid, 21st edition.

Rebrus, P. and Törkenczy, M. (2009). Covert and overt defectiveness in paradigms. In Rice, C. and Blaho, S., editors, *Modeling ungrammaticality in Optimality Theory*, pages 195–236. Equinox, London.

Redington, M., Chater, N., and Finch, S. (1998). Distributional information: A powerful cue for acquiring syntactic categories. *Cognitive Science*, 22(4):425–469.

Regel, S., Opitz, A., Müller, G., and Friederici, A. D. (2015). The past tense debate revisited: Electrophysiological evidence for subregularities of irregular verb inflection. *Journal of Cognitive Neuroscience*, 27(9):1870–1885.

Rescorla, L. A. (1980). Overextension in early language development. *Journal of Child Language*, 7(2):321–335.

Rice, C. (2005). Optimal gaps in optimal paradigms. *Catalan Journal of Linguistics*, 4:155–170.

Rice, C. and Blaho, S. (2009). Modeling ungrammaticality. In Rice, C. and Blaho, S., editors, *Modeling ungrammaticality in Optimality Theory*, pages 1–16. Equinox, London.

Rissanen, J. (1978). Modeling by shortest data description. *Automatica*, 14(5):465–471.

Rivest, R. (1976). On self-organizing sequential search heuristics. *Communications of the ACM*, 2:63–67.

Roberts, I. (2012). Macroparameters and minimalism: A programme for comparative research. In Galves, C., Cyrino, S., Lopes, R., Sandalo, F., and Avelar, J., editors, *Parameter theory and linguistic change*, pages 320–335. Oxford University Press, Oxford.

Rubenstein, H. and Goodenough, J. B. (1965). Contextual correlates of synonymy. *Communications of the ACM*, 8(10):627–633.

Rumelhart, D. E. and McClelland, J. L. (1986). On learning the past tenses of English verbs. In McClelland, J. L., Rumelhart, D. E., and the PDP Research Group, editors, *Parallel distributed processing: Explorations into the microstructure of cognition. Volume 2: Psychological and biological models*, pages 216–271. MIT Press, Cambridge, MA.

Saffran, J. R., Aslin, R. N., and Newport, E. (1996). Statistical learning by by 8-month-old infants. *Science*, 274:1926–1928.

Sag, I. A. (2010). English filler-gap constructions. *Language*, 86(3):486–545.

Sakas, W. G. and Fodor, J. D. (2012). Disambiguating syntactic triggers. *Language Acquisition*, 19(2):83–143.

Sakas, W. G., Yang, C., and Berwick, R. (2016). Parameter setting is feasible. *Theoretical Linguistics*, (To appear).

Salkoff, M. (1983). Bees are swarming in the garden: a systematic synchronic study of productivity. *Language*, 59(2):288–346.

Sankoff, G. and Blondeau, H. (2007). Language change across the lifespan: /r/ in Montreal French. *Language*, 83(3):560–588.

Sapir, E. (1928). *Language: An introduction to the study of speech*. Harcourt Brace, New York.

Scarborough, D. L., Cortese, C., and Scarborough, H. S. (1977). Frequency and repetition effects in lexical memory. *Journal of Experimental Psychology: Human Perception and Performance*, 7(1):3–12.

Schieffelin, B. B. and Ochs, E. (1986). *Language socialization across cultures*. Cambridge University Press, Cambridge.

Schlesinger, I. M. (1971). Production of utterances and language acquisition. In Slobin, D., editor, *The ontogenesis of grammar*, pages 63–101. Academic Press, New York.

Schmid, H. (1995). Improvements in part-of-speech tagging with an application to German. In *Proceedings of the ACL SIGDAT-Workshop*, pages 47–50.

Scholes, R. J. (1966). *Phonotactic grammaticality*. Mouton, Berlin.

Schuler, K., Yang, C., and Newport, E. (2016). Testing the Tolerance Principle: Children form productive rules when it is more computationally efficient to do so. In *The 38th Cognitive Society Annual Meeting*, Philadelphia, PA.

Schütze, C. T. (2001). On the nature of default case. *Syntax*, 4(3):205–238.

Schütze, C. T. (2005). Thinking about what we are asking speakers to do. In Kepser, S. and Reis, M., editors, *Linguistic evidence: Empirical, theoretical, and computational perspectives*, pages 457–485. Mouton de Gruyter, Berlin.

Schütze, C. T. (2011). Linguistic evidence and grammatical theory. *Wiley Interdisciplinary Reviews: Cognitive Science*, 2(2):206–221.

Schütze, C. T. R. (1997). *INFL in child and adult language: Agreement, case and licensing*. PhD thesis, MIT.

Sebastián, N., Cuetos, F., Martí, A., and Carreiras, M. (2000). *LEXESP: Léxico informatizado del español*. Edicions de la Universitat de Barcelona, Barcelona.

Seidl, A. and Buckley, E. (2005). On the learning of arbitrary phonological rules. *Language Learning and Development*, 1(3–4):289–316.

Sharoff, S. (2005). Methods and tools for development of the Russian Reference Corpus. In Archer, D., Wilson, A., and Rayson, P., editors, *Corpus linguistics around the world*, pages 167–180. Rodopi, Amsterdam.

Shepard, R. N. (1987). Toward a universal law of generalization for psychological science. *Science*, 237(4820):1317–1323.

Shi, R. and Melançon, A. (2010). Syntactic categorization in French-learning infants. *Infancy*, 15(5):517–533.

Shipley, E. F., Smith, C. S., and Gleitman, L. R. (1969). A study in the acquisition of language: Free responses to commands. *Language*, 45(2):322–342.

Shlonsky, U. (2009). Hebrew as a partial null-subject language. *Studia Linguistica*, 63(1):133–157.

Siegel, D. C. (1974). *Topics in English morphology*. PhD thesis, MIT. Published by Garland, New York, 1979.

Sigurjónsdóttir, H. (2002). Case marking in child language: How do Icelandic children learn to use case? Master's thesis, University of Iceland, Reykjavik.

Sims, A. D. (2006). *Minding the gap: Inflectional defectiveness in a paradigmatic theory*. PhD thesis, Ohio State University.

Sinclair, J. (1987). *Collins COBUILD, Collins Birgmingham University International Language Database: English language dictionary*. Collins, London.

Sirts, K. and Goldwater, S. (2013). Minimally-supervised morphological segmentation using adaptor grammars. *Transactions of the Association for Computational Linguistics*, 1:255–266.

Skinner, B. F. (1957). *Verbal behavior*. Appleton Century Crofts, New York.

Skousen, R., Lonsdale, D., and Parkinson, D. B., editors (2002). *Analogical modeling: An exemplar-based approach to language*. John Benjamins, Amsterdam.

Sleator, D. D. and Tarjan, R. E. (1985a). Amortized efficiency of list update and paging rules. *Communications of the ACM*, 28(2):202–208.

Sleator, D. D. and Tarjan, R. E. (1985b). Self-adjusting binary search trees. *Journal of the ACM*, 32(3):652–686.

Slobin, D. I. (1971). On the learning of morphological rules: A reply to Palermo and Eberhart. In Slobin, D. I., editor, *The ontogenesis of grammar*. Academic Press, New York.

Sluijter, A. and van Heven, V. (1996). Spectral balance as an acoustic correlate of linguistic stress. *Journal of the Acoustical Society of America*, 100(4):2471–2485.

Smith, H. (1994). Dative Sickness in Germanic. *Natural Language and Linguistic Theory*, 12(4):675–736.

Smith, L. and Yu, C. (2008). Infants rapidly learn word-referent mappings via cross-situational statistics. *Cognition*, 106(3):1558–1568.

Smith, N. V. and Tsimpli, I.-M. (1995). *The mind of a savant: Language learning and modularity*. Blackwell Publishing, Oxford.

Smolensky, P. (1996). The initial state and "richness of the base" in Optimality Theory. Technical Report JHU-CogSci-96-4, Department of Cognitive Science, Johns Hopkins University.

Smolka, E., Zwitserlood, P., and Rösler, F. (2007). Stem access in regular and irregular inflection: Evidence from German participles. *Journal of Memory and Language*, 57(3):325–347.

Snyder, W. (2007). *Child language: The parametric approach*. Oxford University Press, Oxford.

Snyder, W. and Stromswold, K. (1997). The structure and acquisition of English dative constructions. *Linguistic Inquiry*, 28(2):281–317.

Sober, E. (1975). *Simplicity*. Oxford University Press, New York.

Solomyak, O. and Marantz, A. (2010). Evidence for early morphological decomposition in visual word recognition. *Journal of Cognitive Neuroscience*, 22(9):2042–2057.

Sonnenstuhl, I. and Huth, A. (2002). Processing and representation of German *-n* plurals: A dual mechanism approach. *Brain and Language*, 81(1–3):276–290.

Stefanowitsch, A. (2008). Negative entrenchment: A usage-based approach to negative evidence. *Cognitive Linguistics*, 19(3):513–531.

Steriade, D. (1997). Lexical conservatism. In *Linguistics in the morning calm: Selected Papers from SICOL 1997*, pages 157–179. Hanshin, Seoul.

Sternberg, S. (1969). Memory-scanning: Mental processes revealed by reaction-time experiments. *American Scientist*, 57(4):421–457.

Stevens, J., Trueswell, J., Yang, C., and Gleitman, L. (2016). The pursuit of word meanings. *Cognitive Science*, (To appear).

Stockall, L. and Marantz, A. (2006). A single route, full decomposition model of morphological complexity: MEG evidence. *The Mental Lexicon*, 1(1):85–123.

Studdert-Kennedy, M. (1998). The particulate origins of language generativity: from syllable to gesture. In Hurford, J., Studdert-Kennedy, M., and Knight, C., editors, *Approaches to the Evolution of Language*, pages 202–221. Cambridge University Press, Cambridge.

Stump, G. (2001). *Inflectional morphology: A theory of paradigm structure*. Cambridge University Press, Cambridge.

Stvan, L. S. (1998). *The semantics and pragmatics of bare singular noun phrases*. PhD thesis, Northwestern University.

Sweet, H. (1892). *A new English grammar: logical and historical. 1. Introduction, phonology and accidence*. Clarendon Press, Oxford.

Swinney, D. and Cutler, A. (1979). The access and processing of idiomatic expressions. *Journal of Verbal Learning and Verbal Behavior*, 18(5):523–534.

Szagun, G. (2001). Learning different regularities: The acquisition of noun plurals by German-speaking children. *First Language*, 21:109–141.

Szagun, G. (2004). Learning by ear: on the acquisition of case and gender marking by German-speaking children with normal hearing and with cochlear implants. *Journal of Child Language*, 31(1):1–30.

Tabossi, P., Fanari, R., and Wolf, K. (2009). Why are idioms recognized fast? *Memory & Cognition*, 37(4):529–540.

Taft, M. (1979). Recognition of affixed words and the word frequency effect. *Memory and Cognition*, 7(4):263–272.

Taft, M. (2004). Morphological decomposition and the reverse base frequency effect. *The Quarterly Journal of Experimental Psychology*, 57A(4):745–765.

Taft, M. and Forster, K. I. (1975). Lexical storage and retrieval of prefixed words. *Journal of Verbal Learning and Verbal Behavior*, 14(6):638–647.

Taft, M. and Forster, K. I. (1976). Lexical storage and retrieval of polymorphemic and polysyllabic words. *Journal of Verbal Learning and Verbal Behavior*, 15(6):607–620.

Tardif, T., Shatz, M., and Naigles, L. (1997). Caregiver speech and children's use of nouns versus verbs: A comparison of English, Italian, and Mandarin. *Journal of Child Language*, 24(3):535–565.

Tattersall, I. (2012). *Masters of the planet: The search for our human origins*. Macmillan, New York.

Taylor, A. (1994). Variation in past tense formation in the history of English. In Izvorski, R., Meyerhoff, M., Reynolds, B., and Tredinnick, V., editors, *Penn Working Papers in Linguistics 1*, pages 143–158. Penn Linguistics Club, Philadelphia.

Taylor, J. R. (2003). *Linguistic categorization*. Oxford University Press, Oxford.

Tenenbaum, J. B. and Griffiths, T. L. (2001). Generalization, similarity and bayesian inference. *Behavioral and Brain Sciences*, 24(4):629–640.

Terrace, H. S., Petitto, L.-A., Sanders, R. J., and Bever, T. G. (1979). Can an ape create a sentence? *Science*, 206(4421):891–902.

Tesar, B. and Smolensky, P. (2001). *Learnability in Optimality Theory*. MIT Press, Cambridge.

Tettamanti, M., Alkadhi, H., Moro, A., Perani, D., Kollias, S., and Weniger, D. (2004). Neural correlates for the acquisition of natural language syntax. *NeuroImage*, 17:700–709.

Thráinsson, H. (2005). *Íslensk tunga III: Setningar*. Almenna bókafélagið, Reykjavík.

Tomasello, M. (1992). *First verbs: A case study of early grammatical development*. Harvard University Press, Cambridge.

Tomasello, M. (2000a). Do young children have adult syntactic competence? *Cognition*, 74(3):209–253.

Tomasello, M. (2000b). First steps toward a usage-based theory of language acquisition. *Cognitive linguistics*, 11(1/2):61–82.

Tomasello, M. (2003). *Constructing a language*. Harvard University Press, Cambridge, MA.

Topping, D. M. (1973). *Chamorro reference grammar*. University of Hawaii Press, Honolulu. reprinted in 1980.

Trabasso, T. and Bower, G. H. (1975). *Attention in learning: Theory and research*. Krieger, Malabar, FL.

Trammell, R. L. (1978). The psychological reality of underlying forms and rules for stress. *Journal of Psycholinguistic Research*, 7(2):79–94.

Trueswell, J. C. (1996). The role of lexical frequency in syntactic ambiguity resolution. *Journal of Memory and Language*, 35:566–585.

Tulving, E. (1972). Episodic and semantic memory. In Tulving, E. and Donaldson, W., editors, *Organization of memory*, pages 381–402. Academic Press, New York.

Twain, M. (1880). *A tramp abroad*. American Publishing Company, Hartford, CT.

Tyler, A. and Nagy, W. (1989). The acquisition of English derivational morphology. *Journal of Memory and Language*, 28(6):649–667.

Valian, V. (1986). Syntactic categories in the speech of young children. *Developmental Psychology*, 22(4):562.

Valian, V. (1990). Null subjects: A problem for parameter-setting models of language acquisition. *Cognition*, 35(2):105–122.

Valian, V. (1991). Syntactic subjects in the early speech of American and Italian children. *Cognition*, 40(1):21–81.

Valian, V., Solt, S., and Stewart, J. (2009). Abstract categories or limited-scope formulae? The case of children's determiners. *Journal of Child Language*, 36(4):743–778.

Valiant, L. G. (1984). A theory of the learnable. *Communications of the ACM*, 27(11):1134–1142.

Van Lancker, D., Canter, G. J., and Terbeek, D. (1981). Disambiguation of ditropic sentences: Acoustic and phonetic cues. *Journal of Speech, Language, and Hearing Research*, 24(3):330–335.

Van Marle, J. (1992). The relationship between morphological productivity and frequency: A comment on Baayen's performance-oriented conception of morphological productivity. In Booij, G. and van Marle, J., editors, *Yearbook of Morphology 1991*, pages 151–163. Springer Netherlands, Amsterdam.

Vapnik, V. (2000). *The nature of statistical learning theory*. Springer, Berlin.

Vihman, M. M., DePaolis, R. A., and Davis, B. L. (1998). Is there a "trochaic bias" in early word learning? evidence from infant production in English and French. *Child Development*, 69(4):935–949.

Vollmann, R., Sedlak, M., Müller, B., and Vassilakou, M. (1997). Early verb inflection and noun plural formation in four Austrian children: the demarcation of phases and interindividual variation. *Papers and Studies in Contrastive Linguistics*, 33:59–78.

Walker, J. (1936). *Walker's rhyming dictionary*. Dutton, New York.

Wallenberg, J. C., Ingason, A. K., Sigurðsson, E. F., and Rögnvaldsson, E. (2011). Icelandic parsed historical corpus (icepahc). *Version 0.9. Size*, 1.

Wang, Q., Lillo-Martin, D., Best, C. T., and Levitt, A. (1992). Null subject versus null object: Some evidence from the acquisition of Chinese and English. *Language Acquisition*, 2(3):221–254.

Wang, W. S.-Y. (1969). Competing changes as a cause of residue. *Language*, 45(1):9–25.

Wang, W. S.-Y. (1979). Language change: A lexical perspective. *Annual Review of Anthropology*, 8:353–371.

Weide, R. (2008). The CMU pronunciation dictionary 0.7. Carnegie Mellon University.

Weinreich, U., Labov, W., and Herzog, M. (1968). Empirical foundations for a theory of language change. In Lehmann, W., editor, *Directions for historical linguistics: A symposium*, pages 95–195. University of Texas Press, Austin.

Weir, R. H. (1962). *Language in the crib*. Mouton, Berlin.

Werker, J. F. and Tees, R. C. (1984). Cross-language speech perception: Evidence for perceptual reorganization during the first year of life. *Infant Behavior and Development*, 7(1):49–63.

Westermann, G. (2000). A constructivist dual-representation model of verb inflection. In Gleitman, L. and Joshi, A., editors, *Proceedings of the Twenty-Second Annual Conference of the Cognitive Science Society*, pages 977–982. Lawrence Erlbaum, Mahwah, NJ.

Westfal, S. (1956). *A study in Polish morphology: The genitive singular masculine*. Mouton, 's-Gravenhage.

Wexler, K. and Culicover, P. (1980). *Formal principles of language acquisition*. MIT Press, Cambridge, MA.

Weyerts, H. and Clahsen, H. (1994). Netzwerke und symbolische regeln im spracherwerb: experimentelle ergebnisse zur entwicklung der flexionsmorphologie. *Linguistische Berichte*, 154(4):430–460.

Wiese, R. (1996). *The phonology of German*. Clarendon, Oxford.

Wiese, R. (1999). On default rules and other rules. *Behavioral and Brain Sciences*, 22(6):1043–1044.

Wolf, M. and McCarthy, J. J. (2009). Less than zero: Correspondence and the null output. In Rice, C. and Blaho, S., editors, *Modeling ungrammaticality in Optimality Theory*, pages 17–66. Equinox, London.

Wray, A. (1998). Protolanguage as a holistic system for social interaction. *Language & communication*, 18(1):47–67.

Wu, J., Cheng, Z., and Pan, S. (2014). *New Age Chinese-English Dictionary*. Commerical Press, Beijing.

Wunderlich, D. (1999). German noun plural reconsidered. *Behavioral and Brain Sciences*, 22:1044–1045.

Xu, F. and Pinker, S. (1995). Weird past tense forms. *Journal of Child Language*, 22(3):531–556.

Xu, F. and Tenenbaum, J. B. (2007). Word learning as Bayesian inference. *Psychological Review*, 114(2):245.

Yamamoto, K. and Keil, F. (2000). The acquisition of Japanese numeral classifiers: Linkage between grammatical forms and conceptual categories. *Journal of East Asian Linguistics*, 9(4):379–409.

Yang, C. (2000). Dig-dug, think-thunk. *London Review of Books*, 22(16):42–43.

Yang, C. (2002). *Knowledge and learning in natural language*. Oxford University Press, Oxford.

Yang, C. (2004). Universal grammar, statistics or both? *Trends in Cognitive Sciences*, 8(10):451–456.

Yang, C. (2006a). *The infinite gift: How children learn and unlearn the languages of the world*. Scribner, New York.

Yang, C. (2006b). A stochastic model of morphological change. Ms., Yale University.

Yang, C. (2012). Computational models of syntactic acquisition. *Wiley Interdisciplinary Reviews: Cognitive Science*, 3(2):205–213.

Yang, C. (2013a). Ontogeny and phylogeny of language. *Proceedings of the National Academy of Sciences*, 110(16):6324–6327

Yang, C. (2013b). Who's afraid of george kingsley zipf? or: Do children and chimps have language? *Significance*, 10(6):29–34.

Yang, C. (2015). Negative knowledge from positive evidence. *Language*, 91(4):938–953.

Yang, C. (2016). Rage against the machine: Evaluation metrics in the 21st century. *Language Acquisition*, To appear.

Yang, C., Ellman, A., and Legate, J. A. (2015). Input and its structural description. In Ott, D. and Gallego, A., editors, *50th anniversary of Noam Chomsky's Aspects of the Theory of Syntax*. MITWPL.

Yip, K. and Sussman, G. J. (1997). Sparse representations for fast, one-shot learning. In *Proceedings of the National Conference on Artificial Intelligence*, pages 521–527.

Yip, M., Maling, J., and Jackendoff, R. S. (1987). Case in tiers. *Language*, pages 217–250.

Zaenen, A., Maling, J., and Thráinsson, H. (1985). Case and grammatical functions: The Icelandic passive. *Natural Language and Linguistic Theory*, 3(4):441–483.

Zaliznjak, A. A., editor (1977). *Grammatičeskij slovar' russkogo jazyka: Slovizmenenie*. Russkij jazyk, Moskva.

Zhao, Q. and Marcus, M. (2009). A simple unsupervised learner for pos disambiguation rules given only a minimal lexicon. In *Proceedings of the 2009 Conference on Empirical Methods in*

Natural Language Processing: Volume 2, pages 688–697. Association for Computational Linguistics.

Zipf, G. K. (1949). *Human behavior and the principle of least effort: An introduction to human ecology.* Addison-Wesley, Cambridge, MA.

Index

a-adjectives, 180-190
  See also indirect negative evidence
  empirical distribution, 181-184
  morphological structure, 185, 188
  positive evidence, 184-189
  statistical preemption, 181, 183
"Abe", 89-90, 90n,
"Adam", 45, 87-89, 91, 225-226
Akhtar, Nameera, 193
Albright, Adam, 39, 42, 47, 83, 141, 144, 150-152
Alegre, Maria, 43n
Allen, Shanley, 35
Anderson, John, 218
Anderson, Stephen, 49, 107, 113, 141, 159
Anderwald, Lieselotte, 158-160
Angluin, Dana, 174
Aronoff, Mark, 10, 27n, 46, 49, 106-107, 113, 114n, 120, 217
Anshen, Frank, 27n, 64, 120
Aslin, Richard, 96, 179

Baayen, Harald, 38-39, 43n, 47, 55, 64, 107, 108n
Baerman, Matthew, 12, 139, 142
Baker, Mark, 205
Baker's Paradox, 12, 173, 190, 206, 210
Bakovic, Eric, 78
Balota, David, 61
Baronian, Luc, 140-141, 151, 153
Bartke, Susanne, 129
Basque, 23
Bauer, Laurie, 39, 41
Berko, Jean, 11, 31-32, 85-86
Berwick, Robert, 6, 42, 80, 174-175, 218
bilingualism, 137
Bittner, Dagmar, 128, 130
Bloch, Bernard, 26, 44
Blondeau, Hĩl'lene, 224
Bolhuis, Johan, 1
Botha, Rudolf, 39
Bowerman, Melissa, 33, 118, 175, 192, 197n
Boyd, Jeremy, 181, 183
Boyé, Gilles, 152
Braine, Martin, 16, 175
British National Corpus (BNC), 143-145
Bromberger, Sylvain, 222
Brooks, Patricia, 196
Brown Corpus (US English), 16, 19, 23
Brown, Roger, 16, 91, 187
Brown, Dunstan, 49, 73
Bruening, Benjamin, 184-186, 193, 205
Buckley, Eugene, 222
Bush, Robert, 225,
Bybee, Joan, 27n, 38, 83, 86, 149

Caramazza, Alfonso, 53, 55
Carey, Susan, 178
Carnie, Andrew, 194
Catalan, 23
CELEX Corpus (German), 81, 124
Chamorro, 174, 194
Chan, Erwin, 21-24, 42n, 206
Charniak, Eugene, 24
CHILDES, 19-21, 81, 125, 155
Chinese, 8, 59, 108, 195, 219
Chomsky, Noam, 1, 3, 16, 27, 41, 75, 80, 93, 98, 117, 169, 171, 174, 179, 215, 220, 224
Chung, Sandra, 174, 194
Cinque, Guglielmo, 180
Clahsen, Harald, 35-37, 57-58, 121, 349
Clark, Eve, 68n, 118, 181
CMU Pronunciation Dictionary, 98
Coady, Jeffrey, 179
Cohn, Abby, 8, 134n
Coleman, John, 179
Collins, Michael, 24-26
Collins Corpus (COBUILD), 81
Conwell, Erin, 193
Corbett, Greville, 73, 139
Corpus of Contemporary American English (COCA), 143-145
Corpus of Historical American English (COHA), 158-159
Crain, Stephen, 5, 173, 191, 218
Culicover, Peter, 4, 6, 80, 181
Cutler, Anne, 59, 101, 102n
Czech, 23

Dąbrowska, Ewa, 38, 154-166
dative construction, 12
  See also Baker's Paradox; indirect negative evidence
  crosslinguistic variation, 13, 194-195
  empirical distribution, 199-200
  learnability of, 200-212
  lexical arbitrariness, 12, 194-195
  overgeneralization, 192-193, 196, 205
  productivity, 191-193
  restrictions, 190, 193, 197, 206, 210-212
  semantic conditions, 193-196, 200-205, 211
defective gaps, 38, 140-156
  See also Distributed Morphology; Dual-Route Morphology; Network Morphology; Optimality Theory; Paradigm Function Morphology
  as structurally based, 140
  competition-based approaches, 38, 141-142
  condition on, 142
  lexical conservatism, 140-141

Deen, Kamil Ud, 35
Demuth, Katherine, 35, 100, 193
Distributed Morphology, 14, 27, 38, 49, 107, 133, 139, 141
Dresher, Elan, 96
Dressler, Wolfgang, 123, 129
Dual-Route Morphology, 151, 154
See also Clahsen, Harald; English, past tense; Pinker, Steven

Eddington, David, 36, 147, 150n
Ellman, Allison, 90
Elman, Jeffrey, 225
Elsen, Hilke, 124, 126, 128
Elsewhere Condition
See also frequency effects
as linguistic principle, 49-50, 216
as processing model, 50-60
domain generality, 217-218
exceptions before rules, 52-60
in morphosyntax, 164
linear search, 50-51
Embick, David, 107, 141, 144
English Lexicon Project (ELP), 51, 54-55, 98, 105
English
See also stress
U-shaped learning, 11, 87-88
-age, 117
-er, 79, 118-119
-ity, 120-121
-ment, 116
-ness, 112-115
-th, 117-118
nominalization, 10, 46-47, 106-121
past tense, 26-31, 81-93
plural, 91-93
statistical parsing, 24-25
"Eve", 90
evolution of language, 1-2, 15, 215, 223, 227
exception, 3, 7
See also Elsewhere Condition; English, past tense; Tolerance Principle
classifier, 58-59
idiom, 59-60
as rule-governed, 27, 30, 44
Eythórsson, Thórhallur, 161-166

Fenson, Larry, 126, 226
Fikkert, Paula, 11, 96, 100
Finnish, 23
Fleischhauer, Elisabeth, 57-58
Frank, Michael, 177
frequency effects
See also English, past tense; vocabulary size

in acquisition, 28-33
in change, 164
in processing, 52-62, 128
Fodor, Janet Dean, 5, 8, 175, 191
Fongbe, 195
Forster, Kenneth, 50, 52, 119
French, 7, 221

Gagliardi, Annie, 76
Gawlitzek-Maiwald, Ira, 128, 135
Gerken, LouAnn, 100, 110, 187
German, 37, 57-58, 121-136
  gender, 126-129, 131-133
  noun plurals, 58, 121-136
  past participles, 35-36, 57-58, 80-81
  phonology, 129-131
Gildea, Daniel, 224
Gleitman, Lila, 16, 39, 171, 177, 179
Gold, E. Mark, 175
Goldberg, Adele, 4, 137, 181, 186, 193
Goodman, Nelson, 171
Gordon, Peter, 43n
Gorman, Kyle, 43n, 51n, 64, 147, 179n
Greek, 23, 195
Green, Georgia, 192
Griffiths, Thomas, 174
Grimshaw, Jane, 196
Gropen, Jesse, 192-194, 206
Guasti, Maria, 35
Guion, Susan, 11, 102
Gutmann, Sam, 63

Hale, Kenneth, 193
Halldórsson, Halldór, 165
Halle, Morris, 3, 11, 13-14, 27, 44, 49, 93-96, 98, 107, 139-141, 144, 152, 169, 179, 217, 222
Harley, Heidi, 111n, 114n, 194, 200, 205
Harris, James, 36, 147-148, 150n
Harris, Zellig, 96, 110
Hart, Betty, 13, 22, 71, 224
Hauser, Marc, 1-2
Hay, Jen, 38
Hayes, Bruce, 39, 42, 83, 93, 150, 179
Hebrew, 23, 195
Herman, Lewis, 34, 86
Herman, Marguerite Shalett, 34, 86
Hetzron, Robert, 139n, 141, 175
Hippisley, Andrew, 49, 73
Horn, Laurence, 217n
Hornstein, Norbert, 1
Hu, Qian, 59
Hudson, Richard, 14, 141, 142
Hudson Kam, Carla, 68

Hurford, James, 15
Huttenlocher, Janellen, 83, 179, 209

Icelandic, 160-169, 195
Icelandic Parsed Historical Corpus (IcePaHC), 168-169
indirect negative evidence, 172, 174
  See also Sufficiency Principle, 177-180
  Bayesian inference, 175-177, 181-182
  computational intractability, 175
  empirical problems, 181-184, 212
  overhypothesis, 181-183
  subset problem, 174-17
Ingasson, Anton Karl, 168-169
Italian, 23, 35
Inuktitut, 35

Jackendoff, Ray, 27n, 39, 137, 164, 196, 205
Japanese, 59
Jarmulowicz, Linda, 69, 103, 120
Jelinek, Eloise, 194
Jelinek, Frederick, 198
Johnson, Elizabeth, 86, 96
Johnson, Kathy, 91
Johnson, Mark, 24, 34, 224
Jung, Yeun-Jin, 174, 194
Jusczyk, Peter, 86, 96, 102n, 179
Jósson, Jóhannes Gísli, 160-164

Kehoe, Margaret, 11, 100,
Kelly, Michael, 11, 102
Kemp, Charles, 181, 183
Keyser, Samuel Jay, 193
Kiparsky, Paul, 49, 57, 73, 159, 170
Kãűpcke, Klaus-Michael, 128
Korean, 174, 194
Kottum, Steinar, 154
Krifka, Manfred, 193
Kroch, Anthony, 6, 13
Kuczaj, Stanley, 27n, 89-91

Laaha, Sabine, 131
Labov, William, 13, 34, 164, 170, 224
Landau, Barbara, 179
language change, 13, 157-169
  See also vocabulary size
  actuation, 164-169
  analogy as productivity, 157-158,
  formal framework, 159-160
  Icelandic case system, 160-170
  snuck, 158-159
Larson, Richard, 185, 205
Legate, Julie Anne, 10, 90, 95, 164, 173, 219
Levelt, Willem, 53

Levin, Beth, 193, 199-212
Lewontin, Richard, 1
LEXESP (Spanish), 149-151
Liberman, Mark, 93, 105
Lidz, Jeffrey, 76
Lieber, Rochelle, 27, 47, 107, 108n
Lightfoot, David, 169-170
Lignos, Constantine, 42n, 43n, 51n, 64, 110, 177, 187, 226

MacWhinney, Brian, 11
Maling, Joan, 164
Malkiel, Yakov, 147, 148
Maslen, Robert, 192
Marantz, Alec, 14, 30, 49, 51n, 53, 107, 111n, 120, 164
Maratsos, Michael, 13, 89-90
Marcus, Gary, 11, 27, 87, 121, 135, 197
Marcus, Mitchell, 25, 226
Mayol, Laia, 37, 149
Mazurkewich, Irene, 193, 206, 211
McClelland, James, 27, 33, 38
Medin, Douglas, 218
Medina, Tamara, 179
Mervis, Carolyn, 91
Mills, Anne, 76, 127, 131n
Miller, George, 18, 60, 179
Miyagawa, Shigeru, 174, 194
Molnar, Raymond, 42-45
Mosteller, Frederick, 225
Myers, Scott, 30, 44

Nagy, William, 69, 71, 103, 120
Naigles, Lettitia, 209
Network Morphology, 73
Newmeyer, Fritz, 7
Newport, Elissa, 68-69, 96, 225
Niyogi, Partha, 6, 176
Nowenstein, Iris, 164, 167
Noyer, Rolf, 133

O'Donnell, Timothy, 33-34
Oehrle, Richard, 172
Optimality Theory, 38, 78, 80, 141, 172
Osherson, Daniel, 175
overirregularization, 28
  See also language change
  absence in child language, 33-34, 83-87
  bring-brang/brung, 82, 85-87
overregularization, 28-31, 192
  See also English, past tense
  and phonological regularity, 30, 44

Paradigm Function Morphology, 151

Park, Tschan-Zin, 121, 128
Pearl, Lisa, 182
Penke, Martina, 35, 128
Penn Treebank, 25
Perfors, Amy, 176, 183
Pesetsky, David, 73, 153, 205
Peters, Ann, 187
Pierrehumbert, Janet, 67, 179
Pinker, Steven, 11, 14, 27, 39, 174, 193, 196-198
Pitler, Emily, 110
Plag, Ingo, 107
Polish, 40, 154-156
Prince, Alan, 27, 33, 38, 78, 93, 141
productivity
   See also vocabulary size; Wug test
   categorical not gradient, 6, 34-40, 156
   core vs. periphery, 4-8, 137, 218-223
   distributional and inductive learning, 41-45, 75, 99-102, 189, 198
   individual variation, 13, 34, 40, 69-71, 86, 167-168, 224
   morphological parsing, 74, 119
   quantitative measurement, 47, 72, 75
   semi-productivity, 38-40
   tokens vs. types, 67-68
   uniformity, 224

Quine, Willard Van Orman, 171

Randall, Janet, 6, 193, 211
Rastle, Kathleen, 79, 119
Rescorla, Leslie, 179
Ringe, Donald, 158-159
Risley, Todd, 13, 22, 71, 224
Rivest, Ronald, 51
Rumelhart, David, 27, 33
Russian National Corpus, 153

Saffran, Jenny, 96
Sag, Ivan, 4, 25
Sakas, William, 5, 175
Salkoff, Morris, 184
Sankoff, Gillian, 224
Santorini, Beatrice, 125n
Sapir, Edward, 2
Scholes, Robert, 179
Schuler, Kathryn, 68-69
Schüze, Carson, 32, 39, 164
Schwarz, Florian, 125n
Seidl, Amanda, 222
Shepard, Roger, 218
Sigurjónsdóttir, Herdís, 164
Sigurðsson, Einar Freyr, 168-169
Sleator, Daniel, 52

Slovene, 23
Sims, Andrea, 139, 148, 152, 153
Smith, Henry, 160
Smith, Linda, 179
Smith, Neil, 173, 218
Smolensky, Paul, 38, 78, 80, 139
Smolka, Eva, 81
Sprouse, Jon, 182
Sober, Elliot, 48
Sound Pattern of English, The, 3, 8, 10, 27, 47-48, 78, 93, 98, 169, 173, 179n, 216
Spanish, 19-21, 23, 34-37, 147-152
Sternberg, Robert, 50
Snyder, William, 13, 221n
Sonnenstuhl, Ingrid, 58, 128
statistics of language, 15-19
   adult and child language comparison, 22-23
   necessity of general rules, 22-26
   saturation, 21-23
   sparsity, 17-26
Stefanowitsch, Anatol, 181
Steriade, Donca, 140
Stevens, Jon, 177
stress
   See also Sound Pattern of English, The
   Dutch, 100
   and lexical category, 93-96, 98, 103
   and morphology, 98, 105
   and phonology development, 96-97
   initial-stress stage, 99-102
   quantity sensitivity, 10-11, 94-96, 102
   statistics of, 10, 99, 101, 105
Stump, Gregory, 49
SUBTLEX-US (US English), 86, 189, 207
Sufficiency Principle, 172, 77-179, 189, 198, 200, 204-205
   See also a-adjective; dative construction; indirect negative evidence
Sussman, Gerry, 42-45
Swedish, 23
Swinney, David, 59
Szagun, Gisela, 126, 128, 131n, 135

Taft, Marcus, 51, 53-55, 119
Tarjan, Robert, 52
Tattersall, Ian, 2012
Taylor, Ann, 158
Tenenbaum, Joshua, 48, 174, 176
Thornton, Rozalind, 173
Thráinsson, Höskuldur, 164, 167
TIGER Corpus (German), 26
Tolerance Principle, 8, 60-66
   as evaluation metric, 47-49, 97, 169, 216
   derivation, 61-66

Index   261

Maximize Productivity, 72, 102, 126
  recursive application, 71-75, 123
  smaller is better, 66-67, 99, 189, 226
  theory neutrality, 75-76, 78
  transformational learning, 79-80, 97
  unreasonable effectiveness, 64-66, 76-78
Tomasello, Michael, 4, 16, 22, 193, 196, 225
Trueswell, John, 39, 177, 179
Tsimpli, Ianthi-Marie, 173, 218
Tyler, Andrea, 69, 103, 120

Universal Grammar
  computational efficiency, 48-49, 215
  Darwin's problem, 1, 15, 215
  domain generality and specificity, 14, 215, 224
  minimum description length, 48
  Plato's problem, 1, 215, 223
  principles and parameters, 3, 5, 172, 220

Valian, Virginia, 6, 8, 225
variational learning, 1, 8
  and transformational learning, 6, 79-80
  and triggering, 6, 80, 180
  domain generality, 225
  parameter setting, 5-8, 172, 218, 221
Vergnaud, Jean-Roger, 10, 93-95, 106
vocabulary size
  See also "Adam"; "Abe"; "Eve"
  and language change, 159, 167
  and productivity, 66-71, 83, 91, 100, 126
  individual variation, 69-71, 167-168, 196
  token frequency in acquisition, 67, 83, 99-101, 125, 169, 204, 206, 208

Wall Street Journal Corpus, 23
Wallenberg, Joel, 168-169
Warlpiri, 195
Weir, Ruth, 110
Westfal, Stanisław, 154
Wexler, Kenneth, 6, 80, 180
Wiese, Richard, 35, 123-125, 129, 132, 134
Wug test, 11, 31-32, 85-86
  problematic use, 33-34
Wunderlich, Dieter, 123, 129

Xhosa, 35
Xu, Fei, 33-36, 82, 176

Yamamoto, Kasumi, 59
Yang, Charles, 1, 5, 29-30, 68-69, 89-90, 96, 110, 122, 159, 166, 173, 177, 180, 183-184, 218-219, 225
Yaqui, 195
Yip, Moira, 164
Yip, Kenneth, 42-45

Zaenen, Annie, 164
Zhao, Qiuye, 226
Zipf's Law, 17-19
  ecological condition on language acquisition, 26, 41, 106, 109, 143, 183, 189, 204, 206, 212, 224, 223-226
  in Tolerance Principle, 61-66, 76-77
  psychological interpretation, 76-78